ENGINEERING & COMPUTER GRAPHICS WORKBOOK

Using SOLIDWORKS 2017

Ronald E. Barr, Ph.D.
Professor

Thomas J. Krueger, Ph.D.
Senior Lecturer

Davor Juricic, D.Sc.
Professor Emeritus

Mechanical Engineering Department
The University of Texas at Austin

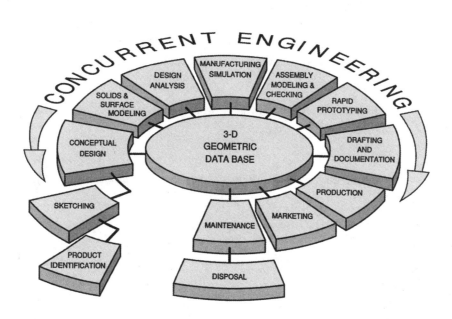

SDC Publications
P.O. Box 1334
Mission, KS 66222
913-262-2664
www.SDCpublications.com
Publisher: Stephen Schroff

The instructions in this Engineering and Computer Graphics Workbook are based on the following release of SOLIDWORKS:

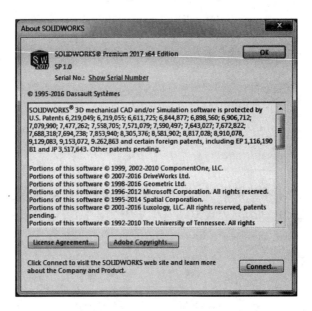

Examination Copies
Books received as examination copies are for review purposes only and may not be made available for student use. Resale of examination copies is prohibited.

Electronic Files
Any electronic files associated with this book are licensed to the original user only. These files may not be transferred to any other party.

ISBN-13: 978-1-63057-057-6
ISBN-10: 1-63057-057-5

Printed and bound in the United States of America.

Table of Contents

NOTES:

Computer Graphics Lab 1:
2-D Computer Sketching I

INTRODUCTION TO SOLIDWORKS

SOLIDWORKS© is a parametric solid modeling package that is used to build solid computer models. The designer starts with a 2-D sketch that at first is loosely defined. Dimensions and other constraints are applied to the 2-D sketch to fully define its geometry. The sketch is turned into a 3-D solid model using the extrusion or revolution command. If the base part needs to be altered, the designer can simply change the dimensions of the original sketch or definition of the base feature, and then rebuild the part with a click of the mouse. Once the base part is defined, new sketches can be defined on planes around the base part, and they can then be extruded to form bosses and cuts on the part. Also, special design features can be easily added to the part, such as fillets, chamfers, and counterbores. This is called feature-based modeling and it is an approach to 3-D computer modeling. In this first computer graphics lab, you will concentrate on 2-D computer sketching methods, and will extrude or revolve simple parts. Four different, short exercises are provided.

SOLIDWORKS TOOLBARS SETUP

Your instructor will show you how to launch SOLIDWORKS on your computer. When you first launch SOLIDWORKS the screen will appear as shown in **Figure 1-1**. Your first task is to set up your SOLIDWORKS screen layout to have a common appearance that will be referenced throughout this workbook. To accomplish this, follow these instructions. Just to the right of the SOLIDWORKS icon at the upper left of the screen, there is an arrow. Click on this arrow to get the basic tab menu. To keep this on the screen, **Select** the **stickpin** at the right end of the menu that appeared. This will always keep this menu visible. Go to the **View** pull-down menu on the top of screen, select **Toolbars** and deactivate all the toolbars except **Dimensions/Relations, Features, Sketch, Standard, Standard Views,** and **View,** as shown in **Figure 1-2**. This will arrange the common toolbars around your screen for easy access during your exercises. Note: these toolbars will not appear until you have begun to construct a new part. As you get familiar with the screen and the commands, you may choose to customize your toolbars to fit your needs.

Figure 1-1. The initial screen in SOLIDWORKS.

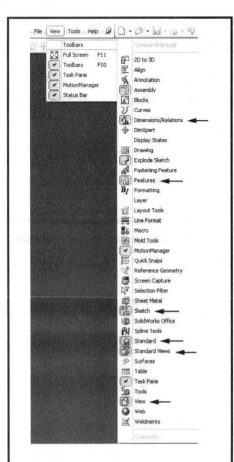

Figure 1-2. Setting the Proper Screen Toolbars during Setup.

STARTING A NEW PART

To start your session with SOLIDWORKS, you need to pull down the **File** main menu and click on the **New** command. This will invoke an on-screen menu, shown in **Figure 1-3**. You have three options: Part, Assembly, and Drawing. To start a new model, click **Part** and then **OK.** Once you have clicked the **OK** button, the screen changes into the SOLIDWORKS computer working space with all the previously checked toolbars ready to be used. Your first step is to study this initial screen layout as depicted in **Figure 1-4**. Some of the toolbars may appear at other positions on the screen.

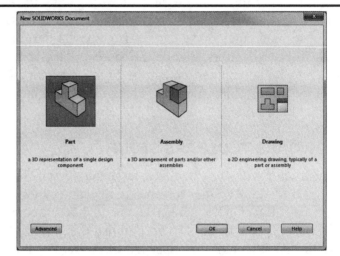

Figure 1-3. Starting a New Part in SOLIDWORKS.

THE SOLIDWORKS SCREEN LAYOUT

The SOLIDWORKS screen layout when you start a part uses the top, left, and right side of the screen to arrange various menus and toolbars. The center of the screen is the computer sketching area for your design work. Study **Figure 1-4** to become familiar with this standard screen layout. Each of these menus and toolbars will be described in the subsequent paragraphs.

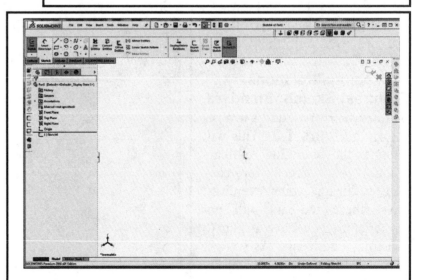

Figure 1-4. The SOLIDWORKS Screen Layout with Major Menus and Toolbars Shown.

MAIN MENU – to keep this menu on the screen, select the stick pen at the right end of the menu bar.

File is where you can create New files and Open, Save, Print Preview, and Print existing files. You also can set up your Page on this menu.

Edit is where you can Undo the last command and Cut, Copy, and Paste Entities. It also has the Rebuild command that you may find convenient when you make changes to your model.

View is where you Redraw and set Display parameters. You can set the View Orientation, and select View commands like Zoom, Rotate, and Pan.

Insert is where the major feature creation commands are located. Here you can create the Base part, Cut the part, and invoke many of the parametric features.

Tools has an assortment of helpful commands. It has Sketching Tools, Dimensioning Tools, and Relations Tools. It also has analysis tools like Mass Properties.

Window is where you can set some window properties like vertical and horizontal tiling.

Help is where you can access on screen listings of the command functions and other information that will help you navigate and learn SOLIDWORKS.

FEATURE MANAGER TREE

SOLIDWORKS has a special "Feature Manager Tree" capability that keeps track of the solid modeling process. It relates each feature with the appropriate sketch in a tree diagram and allows you to edit your feature later if needed. The "Feature Manager" will appear after you have begun a new part or have opened an existing solid model. It is located in the upper left part of the screen. **Figure 1-5** shows an example of the Feature Manager.

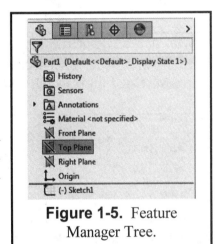

Figure 1-5. Feature Manager Tree.

THE REBUILD BUTTON

The "**Rebuild Button**" is located on the Standard Toolbar and looks like a red/green traffic light, as shown in **Figure 1-6**. When you press it with the mouse button, it will rebuild your model and update any changes that have been made. It terminates any commands that may have been started.

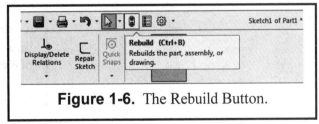

Figure 1-6. The Rebuild Button.

TOOLBARS

Toolbars are clickable icons along the top and side of the screen that permit quick and easy access to some of the more common commands used in SOLIDWORKS. To see all of the Toolbars available to you, move your cursor to the blank gray area above the sketch area and click the Right Mouse Button. You can see the toolbars, grouped by function, in the general screen layout of **Figure 1-4** on page **1-2**.

View and **Standard View Toolbars** are shown in **Figure 1-7**. The View Toolbar contains buttons for several Zoom commands, the Rotate View command, and the Pan (Move) View command. It also has icons to represent your model as a wireframe, with light gray hidden lines, with hidden lines removed, as a shaded image, and as a shaded image with a shadow. The Standard View Toolbar allows you to select Front, Back, Left, Right, Top, Bottom, Isometric, and Normal Views.

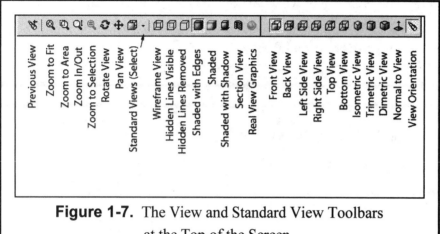

Figure 1-7. The View and Standard View Toolbars at the Top of the Screen.

SKETCH TOOLBAR: The sketch toolbar, shown in **Figure 1-8,** can be divided into sketch entities, relations, and dimension tools. To start, you click on the **Sketch** icon. As you establish 2-D sketch entities, you can use **Dimension** and **Add Relation** buttons to fix the geometric parameters of your design. To aid you in your sketching, numerous 2-D entity icons are available, such as **Line, Arc, Circle, and Rectangle**. To edit your sketch, several common commands like **Trim, Mirror, Fillet**, and **Offset** are available.

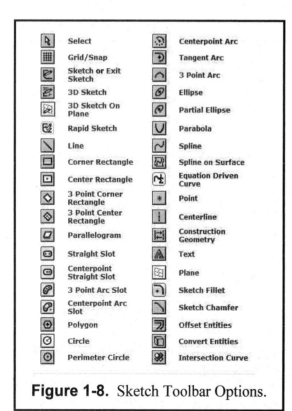

Figure 1-8. Sketch Toolbar Options.

SKETCHING PLANES

Before you start a 2-D sketch with SOLIDWORKS, you must define (highlight) a sketch plane. The three default orthogonal sketch planes are the three standard orthographic view planes: **Front, Top**, and **Right**, as shown in **Figure 1-9**. Later, you can also pick any plane or flat surface on a base part that has been created. The surface does not necessarily have to be parallel to one of these three principal planes.

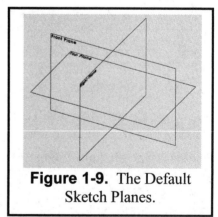

Figure 1-9. The Default Sketch Planes.

LINE COLORS

SOLIDWORKS has several ways to feedback information to the user. One of these ways is to use different line colors. When sketching 2-D profiles, the following color codes prevail.

Cyan lines mean that the entities are **Selected**.
Blue lines mean that the dimensions of the geometry are **Under Defined**.
Black lines mean that the dimensions of the geometry are **Fully Defined**.
Yellow lines mean that the dimensions of the geometry are **Over Defined** (undesirable).
Gray lines mean that these entities are not in the active sketch.

STANDARD TEMPLATES

Before you start any exercises in SOLIDWORKS, it is a good idea to establish some standard drawing templates and title blocks to reflect the two styles of engineering measurements. One set will be for the American National Standards Institute (ANSI) for Inches and the other for Metric.

To start, go to **File,** select **New,** then **Part** and **OK** this selection. Go to the **Tools** pull-down menu. Select **Options**. Under the **Document Properties** tab, make sure that the **Dimensioning Standard** is set for **ANSI** (see **Figure 1-10**). Next, select **Units** in the left-hand column of the menu. In the window that appears, activate **IPS (Inch, Pound, Second)**. You should also increase the Decimal Places to four **(4)** as indicated in **Figure 1-11**. Close the "Document Properties" menu with the **OK** button. Now go to **File** and select **Save As**. First select the **FOLDER** you wish to save this template in, then

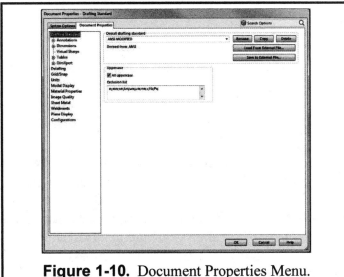

Figure 1-10. Document Properties Menu.

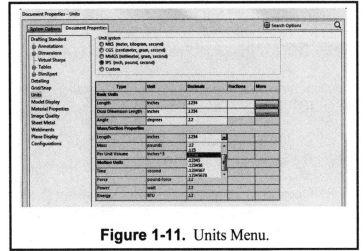

Figure 1-11. Units Menu.

title your part as **ANSI-INCHES,** select **Save as Type** and select **Part Templates (*.prtdot)**. **Make sure that you are saving this into your personal folder.**

Repeat the above process, only this time, the **Units** you select will be **MMGS (Millimeter, Gram, Second)**. Two decimals should be sufficient for metric measurements. Save this template to your **FOLDER**, change your title to **ANSI-METRIC** and select the **Save as Type** to **Part Templates (*.prtdot)**. **Make sure that you are saving this into your folder.**

STANDARD TITLE BLOCK DRAWING SHEETS

The next process is to set up a Title Block to reflect the two measurement styles. You may have to draw a title block (see **Figure 1-12**) if one has not been provided. An example of a title block, with suggested dimensions and style, is available in **Appendix A**. If a Title Block has been provided, go to **File** and **Open** that drawing file. Go to the **Tools** pull-down menu and select **Options**. Under the **Document Properties** tab make sure that the **Dimensioning Standard** is set for **ANSI**. Next, select **Units** in the left-hand column. In the window that appears, activate **IPS (Inch, Pound, Second)**. Three **(3)** decimal places will be all you should use here unless you are doing very high tolerance drawings.

Go to **Insert – Annotations - Note** to add your personal information, such as **NAME, DESK NUMBER**, and **SECTION NUMBER**. In the **Feature Manager Tree**, right click on **Sheet 1** and select **Edit Sheet**. Now that this is done, go to **File** and **Save As**. First select the folder you wish to save this template in, then title your part as **TITLEBLOCK-INCHES** and select the **Save as Type** and select **Drawing Template (*.drwdot)**. **Make sure that you are saving this into your personal folder.**

Repeat the above process, only this time, the **Units** you select will be **MMGS (Millimeter, Gram, Second)**. Two **(2)** decimals should be sufficient for metric measurements. Add your personal information such as **NAME, DESK NUMBER**, and **SECTION NUMBER**. In the **Feature Manager Tree**, right click on **Sheet 1** and select **Edit Sheet**. Save this template in the same way, but change your title to **TITLEBLOCK-METRIC** and select the **Save as Type** and select **Drawing Template (*.drwdot)**. **Make sure that you are saving this into your personal folder.**

INSERTING A PART (RENDERED PICTORIAL) ONTO A TITLE BLOCK SHEET

For each of the assignments you will be asked to submit a printed copy of the part that you built in SOLIDWORKS. The following will be the procedure to accomplish this requirement.

A. Build the solid model according to instructions and save that part in your folder with the file extension of **.sldprt**.

B. Open the appropriate Title Block drawing sheet (see **Figure 1-12** for example). The Title Block drawing sheet that you open should have the same units as the solid part (metric or inches).

C. Pull down **Windows** and then select **Tile Vertically**. This allows you to see all the windows that you have open at the same time.

D. Select the Part Name in the **Feature Manager Tree** and while holding down the left mouse button, drag it onto the Title Block drawing sheet. At this time you have several options available to choose from. For most of your labs you will be choosing a pictorial view or current view. In unit nine and ten you will be choosing the three views layout of the object that you insert onto the Title Block. If you do not see the three views, right click on **Sheet 1** in the **Feature Manager Tree** and select **Edit Sheet**.

E. Pick the window around the pictorial view. You get a menu window that lets you adjust the scale of the image to better fill the sheet.

F. In the drawing area of the Title Block Sheet, **Insert** the title of the object. Provide the exercise number in the upper right-hand box of the Title Block. Your Title should be placed in an open area of the Title Block and a larger font size, i.e., **20 to 24 PT OR .25"**. Your name, desk, Sec and exercise number should be font size **12pt or .125"**.

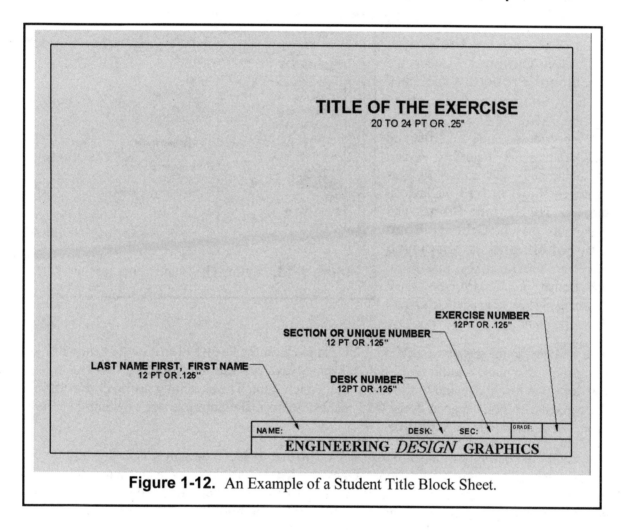

Figure 1-12. An Example of a Student Title Block Sheet.

Exercise 1.1: CARBON FIBER GASKET

In this first exercise, you will build a Carbon Fiber Gasket part. This part is primarily a 2-D object with a thin depth that will be extruded. The gasket features can be drawn with many of the 2-D sketching entities and editing features using the sketch toolbars. Go to your folder and **Open** the file **ANSI-INCHES.prtdot.** In order to avoid corrupting the **ANSI-INCHES.prtdot** template go to **File - Save as:** under **File Name**, type **CARBON FIBER GASKET**, and under **Save as type**, select **.sldprt** and select **SAVE.**

First set up the sketching grid and units. Pull down the **Tools** menu and select **Options**. Under the **Document Properties** tab, click the **Grid/Snap** tab and click **on** (√) all the "Grid" boxes. Set the "Major grid spacing" to **1.00** and "Minor lines per major" to **4**, and set "Snap points per minor" to **1**, as shown in **Figure 1-13**. Click on the **Go To System Snaps** and make sure that the **Grid** and **Snap only when grid is displayed** boxes are checked. Also make sure the **Units** are in **Inches** to **3** decimal places. Then click **OK** to close the menu.

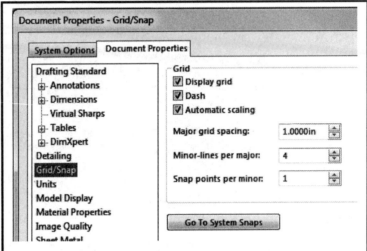

Figure 1-13. Setting Grids and Units for the Sketch.

You will design the gasket in the front view, so click on the **Front Plane** in the Feature Manager tree. The front plane should highlight in cyan. Next, activate the **Sketch Tab** and select the **Sketch** icon just above and to the left of the Sketch Tab. The sketching grid will now appear on the screen with grids spaced every 0.25 inches. Also make sure you are viewing this from the Front View orientation (see **Figure 1-7**).

Now study the outline of the gasket shown in **Figure 1-14**. You will sketch this outline on the quarter-inch grids using the line tool. Click the **Line** icon on the sketch menu toolbar. Now draw five lines according to the specifications in **Figure 1-14** and detailed below. All **(X,Y)** coordinates are relative to the sketch "origin" seen in the middle of the screen. You are able to track your coordinates at the bottom-right of the sketch screen.

> The *top line* starts at **(-3.00, 1.00)** and ends at **(3.00, 1.00)**.
> The *right-side line* starts at **(3.00, 1.00)** and ends at **(3.00, -1.50)**.
> The *bottom right diagonal line* starts at **(3.00, -1.50)** and ends at **(0.00, -3.50)**.
> The *bottom left diagonal line* starts at **(0.00, -3.50)** and ends at **(-3.00, -1.50)**.
> The *left side line* starts at **(-3.00, -1.50)** and ends at **(-3.00, 1.00)**.

The 2-D outline of the gasket should now be finished and completely enclosed.

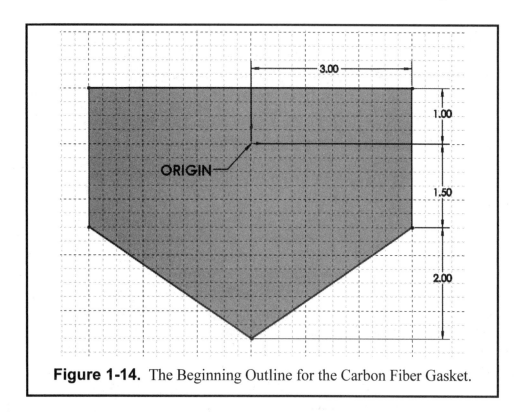

Figure 1-14. The Beginning Outline for the Carbon Fiber Gasket.

The five corners of the gasket are sharp and need to be rounded. This can easily be accomplished using the fillet editing command. Click the **Fillet** icon (round corner symbol) on the sketching toolbar. Notice that a "Sketch Fillet" menu now appears where the Feature Manager tree usually is displayed (**Figure 1-15**). Key in a fillet radius of **0.50** inches. Now, one by one, click on the two edges of the corners (or merely click on the point of intersection) to create a fillet there. Once the five fillets are added, click on the green (√) button on the "Sketch Fillet" menu. The five fillets are now completed as shown later in **Figure 1-16**.

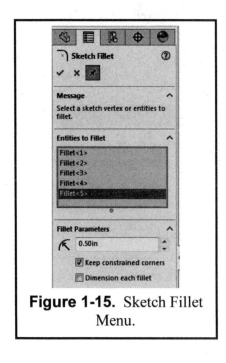

Figure 1-15. Sketch Fillet Menu.

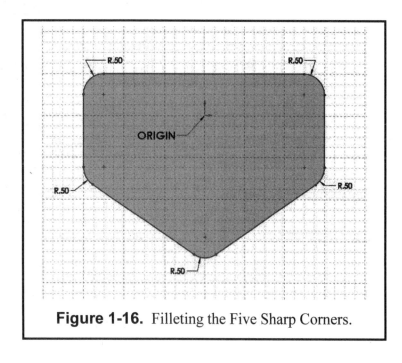

Figure 1-16. Filleting the Five Sharp Corners.

Using the **Straight Slot** sketching tool, draw a slot. Locate the first center at **(-1.5, -1.5)** then drag your cursor to **(1.5, -1.5)**. After you select the second point, move your cursor vertically to form the slot. **Select** the **Smart Dimension** icon and dimension the arc of the slot to be **0.25"** to complete the slot. Next, select the **Circle** sketching tool, and create the five circles concentric to the fillets (rounded corners) of the outline. All circles have the same diameter of **0.375** inches. *Note:* After drawing the five circles use the Smart Dimension to dimension the circles to make sure that they have a diameter of **0.375** inches. **Note:** While you are sketching the lines, rectangle, or circles, the end of the cursor shows a small icon that indicates the current type of sketch entity that you are using. This is called a "smart cursor." A small grid icon shows when the grid snap is activated.

Many solids have symmetrical features and with the following commands those features become very easy to replicate. You will now draw two polygons on the gasket that are symmetrical about a centerline. First select the **Centerline** command under the **Line** menu and draw a vertical centerline through the **Origin**. Next, pull down **Tools**, select **Sketch Tools**, and then select **Dynamic Mirror.** The centerline will now have two small dashes at either end to indicate that it is in the Dynamic Mirror mode. Next, select the "**Polygon**" icon that looks like a hexagon in the sketch menu. A menu will appear on the screen where the previous fillet menu appeared. Draw the polygon on the gasket to the right side of the centerline. First, **locate the center** of the Hexagon. Then **move your cursor horizontally** from that center to locate a vertex of the polygon. Now study the actual parameters needed for the polygon and then set them as shown in **Figure 1-17**.

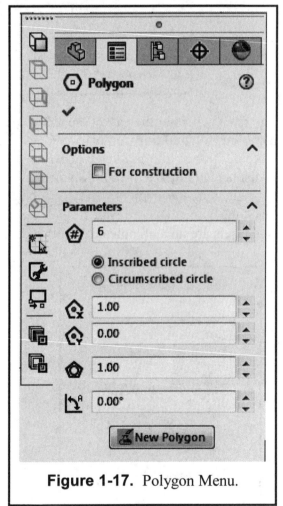

Figure 1-17. Polygon Menu.

    Number of sides = **6**
    Select - Inscribed circle
    X-origin = **1.00**
    Y-origin = **0.00**
    Inscribed circle diameter = **1.00**
    Rotation angle = **0°**

Once the parameters are set, click on the green (√) button on the "Polygon" menu to close the menu and complete the polygons, as shown in **Figure 1-18**. This completes the sketch and can now be extruded to create the solid model.

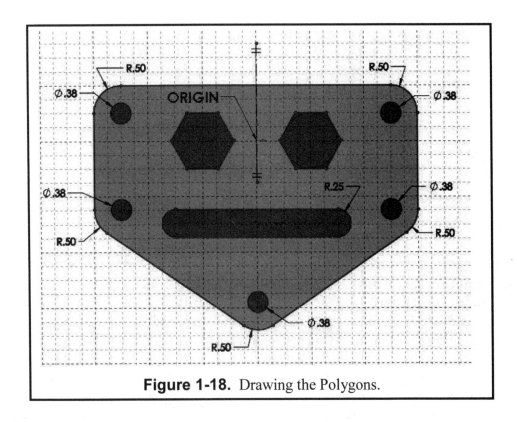

Figure 1-18. Drawing the Polygons.

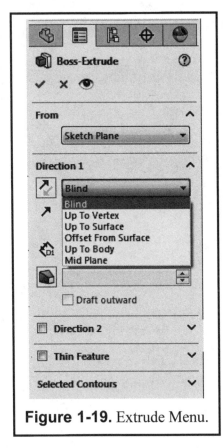

Figure 1-19. Extrude Menu.

EXTRUDING A TWO DIMENSIONAL SKETCH

The **Extruded Boss/Base** function is found in the **Feature** pull down tab. This command will extrude a sketch or selected sketch contours in one or two directions to create a solid feature. Under **Direction 1**, **Blind** is the default selection; however, the down arrow reveals additional choices as shown in **Figure 1-19**. The extrusion options you will be using are **Blind**, **Up to Surface** and **Mid Plane**. The **Blind** option will extrude the pattern to a specified distance. To change the direction of the extrusion, click on the opposing arrow button next to the "**Blind**" box (See **Figure 1-20**). **Up to Surface** will wrap the extrusion to fit the shape of the surface you chose. **Mid Plane** will extrude your pattern equally in both directions from the sketch.

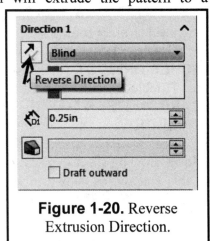

Figure 1-20. Reverse Extrusion Direction.

EXTRUDING THE GASKET THICKNESS

Now extrude the gasket to its proper thickness. Select the **Features** pull-down tab and select **Extruded Boss/Base** or pull down the **Insert** main menu, select **Boss/Base**, and then **Extrude**, as shown in **Figure 1-19**. The "Extrude" menu then appears on the screen in the Feature Manager area (**Figure 1-21**). Set the parameters for the extrusion:

Direction One = **Blind**
Distance1 = **0.25** inches

Notice that the extrusion appears as a preview in the computer sketching area in an Isometric view. Click on (√) the green button on the "Extrude" menu to finish.

Now you can view your model. On the "View Orientation" window click **Trimetric**. You should see the thin solid model appear as a trimetric view (see **Figure 1-22**). Also try **Isometric**, **Dimetric**, and **Front** view orientations. You can experiment with some of the viewing operations in SOLIDWORKS. Click the **Rotate View** icon (it appears as circular arrows on the top toolbars) or press down on the center wheel of the mouse. Now rotate your model into different configurations. Also try several **Zoom** commands and the **Pan** command.

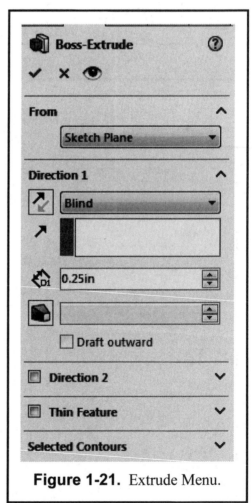

Figure 1-21. Extrude Menu.

If you would like to change the color of your model, click on the model name in the Feature manager tree and then select the colored ball in the menu bar. You can then assign any color you wish to the model.

You should save your model. Pull down the **File** menu and select **Save As**. On the "Save As" menu, select your appropriate file folder in your account, type in the part name **CARBON FIBER GASKET** and the file type as **sldprt**, then click **Save**.

Before you finish this exercise, you should print a hard copy for submission to your instructor. *Check with your instructor first for any special printing instructions.* You may need to use a special sheet like the Title Sheet shown in **Figure 1-12**. If you open **TITLEBLOCK-INCHES.DRWDOT** you must do a **Save As** and save the Titleblock with the name that you want to use **(CARBON FIBER GASKET)** and change the extension to **SLDDRW**. You will now be able to insert the rendered **CARBON FIBER** image into your **Title Block** drawing sheet that was created in the introduction to this chapter, as shown in **Figure 1-22**. To put this object on a Title Sheet, follow the instructions on **Page 1-7**.

If you check (√) **off** the "Use documents Font" in the menu, you can change the font, for example, to:

Font = **Arial**
Font Style = **Regular**
Points = **12**

Place the title of the part in a convenient location of the title sheet. The title should be at 24 pts or .25" in height. Also place the exercise number in the blank to the right of the GRADE box.

Now **SAVE** your drawing sheet to your directory as **CARBON FIBER GASKET.slddrw**. Take note that the title of the part and the drawing are the same but the extension has changed. The solid model and the drawing are linked, meaning that any changes made to the solid model will automatically be updated on the drawing.

Before you print your copy, make sure you have set the **File**, **Page Setup** to **Landscape** mode. Now pull down the **File** and **Print Preview** commands from the top main menu, then the **Print** button to send it to the default printer.

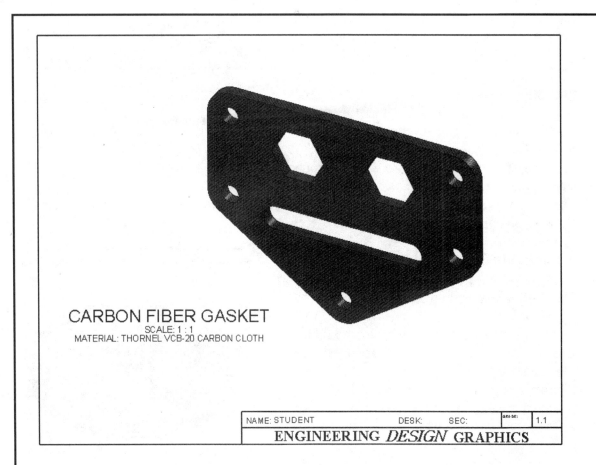

CARBON FIBER GASKET
SCALE: 1 : 1
MATERIAL: THORNEL VCB-20 CARBON CLOTH

| NAME: STUDENT | DESK: | SEC: | GRADE: | 1.1 |

ENGINEERING *DESIGN* GRAPHICS

Figure 1-22. CARBON FIBER GASKET Model Inserted into the Title Block Drawing Sheet.

Exercise 1.2: COVER PLATE

There are many different approaches to creating a 2-D computer sketch using **SOLIDWORKS**. The approach will depend on the design intent and also on the designer's own preferences and choices. In the previous Exercise 1.1, geometry was constructed on a frontal plane using X-Y coordinate geometry and a grid system. In this Exercise 1.2, you will construct the 2-D geometry on a horizontal (top) plane since this is the normal orientation of the cover plate. Also, instead of using a grid coordinate system for laying out the design, you will use simple dimensions related to the geometry and to the computer space origin.

Go to your folder and **Open** the file **ANSI-INCHES.prtdot**. In order to avoid corrupting the **ANSI-INCHES.prtdot template** go to **File - Save as:** under **File Name**, type **COVER PLATE**, and under **Save as type**, Select **.sldprt** and select **SAVE**. Click the **Top plane** in the "Feature Manager." Also click **Top** view in the Standard View toolbar to see the top plane in *cyan* on the screen. Now click the **Sketch** tab and **Select** the **Sketch Icon** on the upper left toolbar, then click the **Circle** icon. Draw a large circle that is centered at the origin. For now, the diameter is unimportant because you will dimension it. Click the **Smart Dimension** icon (slanted dimension line) in the upper left area of the screen. Pick a point on the circumference of the circle and drag the dimension out to the upper right side of the circle. The dimension is not likely to be correct; as soon as you place the dimension a small Modify box appears (see **Figure 1-23**). Key in **10.00** inches and click the green (√) check box. The diameter of the circle changes to the new value. If the circle gets too big, use **Zoom** or Select the **Top View** again to see it better.

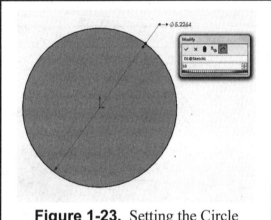

Figure 1-23. Setting the Circle Diameter.

Now you will extrude the circle into the solid base part of the cover plate. Select the **Features** icon and select **Extrude** or pull down **Insert**, select **Boss/Base**, and then **Extrude.** Set the parameters for the extrude to be:

       Direction One = **Blind**
       Distance1 = **0.50** inches

Click the green (√) button on the "Extrude" menu to finish. View the cover plate as an **Isometric** (**Figure 1-24**).

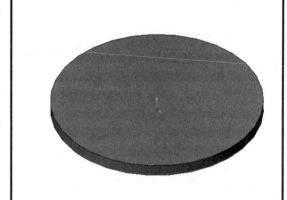

Figure 1-24. Base Part of the Cover Plate.

You will now draw the geometry for the indentation part of the cover plate. **Click on the top surface of the Cover Plate model** - it should turn *blue.* This will be the next sketch surface. Also select **Top** on "View Orientation" for better viewing of the sketch plane. Click the **Sketch Tab** and then select the **Sketch** icon to start the next sketch. Select **Circle** and draw a circle centered at the origin. Now draw a second bigger circle centered at the origin but <u>not</u> as big as the 10.00 inch diameter of the base plate. Now, using the **Smart Dimension** icon, set the diameter of the smaller new circle to **3.50** inches and the diameter of the bigger new circle to **8.50** inches.

Using the **Line** sketching tool, draw a horizontal line from the origin toward the right. Next, draw a line from the origin at some angle above the horizontal line. Both should span across the two circles as shown in **Figure 1-25**. Next, **Dimension** the angle to **22.5 degrees**. Select the horizontal line and then the angled line to change the Smart Dimension into an angular dimension (**Figure 1-25**).

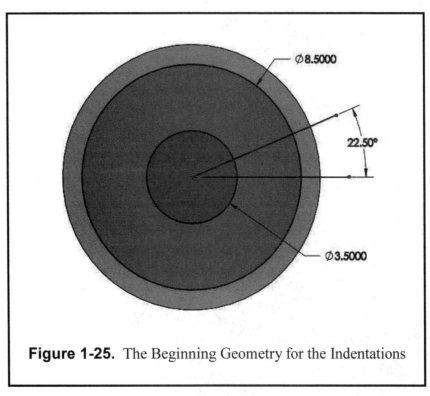

Figure 1-25. The Beginning Geometry for the Indentations

Now select the **Trim** icon on the sketch editing toolbar. In the area where the Feature Manager tree is you will see a **Trim - Options** box. Select the last option (**Trim to Closest**). A scissors icon now appears on the end of the cursor indicating that you are in the trim mode. Trim the line portions (two places) that extend beyond the big circle by just clicking on them with the **LMB**. Next, trim the two line pieces from the origin to the small circle. Finally, trim the portions of the circles that lie outside of the angle. You should now have a pattern like in **Figure 1-26**. Go to the **Linear Sketch Pattern** pull down menu and **Select Circular Sketch Pattern**. Activate the first box and **Select** the **Origin**. Change the number of instances to **8**. The last step is to activate the box titled **Entities to Pattern** and select the four elements of the pattern **(See Figure 1-27)**. This final pattern for the indentations is shown in **Figure 1-28**.

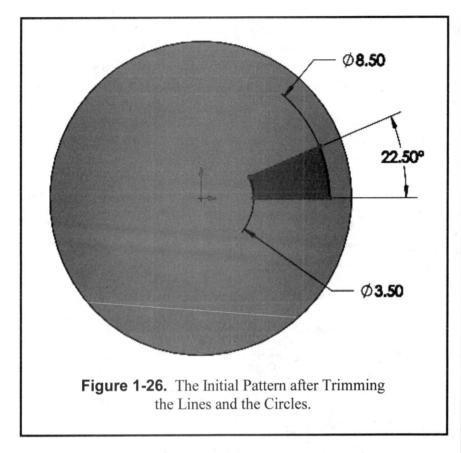

Figure 1-26. The Initial Pattern after Trimming the Lines and the Circles.

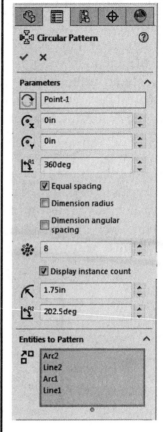

Figure 1-27. The Circular Pattern.

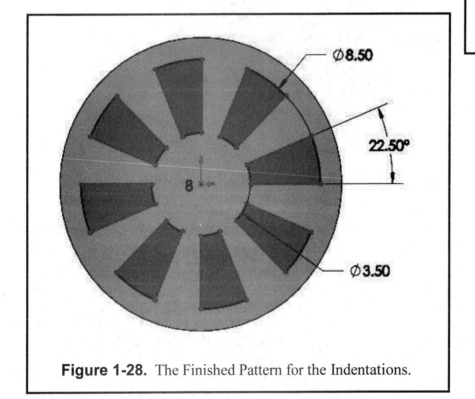

Figure 1-28. The Finished Pattern for the Indentations.

Now you will cut the indentation pattern into the cover plate. It is advisable to **go to an Isometric** view when executing an extrusion of any kind. Select the **Features** icon and select **Extruded Cut**. The "Cut Extrude" menu appears in the Feature Manager area. Key in these parameters:

Direction One = **Blind**
Distance1 = **0.25** inches

Click the green (√) button to finish. View the cover plate as a rotated **Isometric** (**Figure 1-29**).

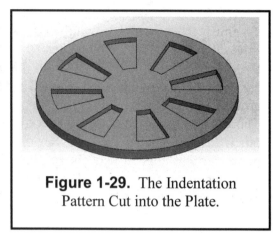

Figure 1-29. The Indentation Pattern Cut into the Plate.

Now create the lift hole in the middle of the plate. Click the cursor on the top center surface of the cover plate (it should turn *blue*). Also, select a **Top** view orientation. You might also need to **Zoom** in to see the center portion better. Click the **Sketch Tab** and then select the **Sketch** icon to get into the sketching mode. Since the lift hole is symmetric about the origin, you can use a mirror function. So start by drawing a vertical **Centerline** (CL) through the origin. Next, go to **TOOLS – SKETCH TOOLS - DYNAMIC MIRROR**. There should be short horizontal lines on either end of the centerline you drew. Every operation you perform on the right side is duplicated on the left side. Now, draw a small **Circle** to the right side of the centerline. Then draw two parallel **Line**s that start at the centerline and go through part of the circle, as indicated in **Figure 1-30**. Now apply the following inch **Dimensions** to the sketch geometry.

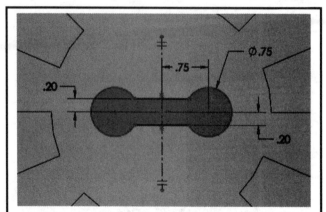

Figure 1-30. Geometry and Dimensions for the Lift Hole.

Circle Diameter = **0.75**
Lateral Dist. from CL = **0.75**
Line1 from origin = **0.20** (above)
Line2 from origin = **0.20** (below)
Circle center = **horizontal with origin**

Now **Trim** the two line pieces that overhang the circle, and then trim the part of the circle between the two parallel lines. If you get a warning from SOLIDWORKS about eliminating associated dimensions while you are trimming, just ignore it and answer **Yes** to proceed with the trim.

Cut the lift-hole pattern through the cover plate. It is advisable to **go to an Isometric view** when executing an extrusion of any kind. Select the **Features** icon and select **Extruded Cut** or pull down **Insert**, select **Cut**, and then **Extrude**. The "Cut Extrude" menu appears in the Feature Manager area. For Direction1, just select **Through All**. Click the green (√) button to finish this extrude-cut operation and view the lift hole in **Figure 1-31**.

To finish the cover plate, you need to add four circles around the perimeter to make the guide holes. Click the cursor on the top surface again and **Sketch** four **Circles** around the perimeter of the cover plate. The circles should be at 90 degrees to each other. The centers should be on the perimeter and aligned with the origin. The diameter of all four circles should be **0.75**, as shown in **Figure 1-31**.

OR

Since you have already used the Circular Sketch Pattern, you could draw a **Circle** with a diameter of **.75** and use the **Circular Sketch Pattern** with the center at the **Origin** and the instances set at **4**.

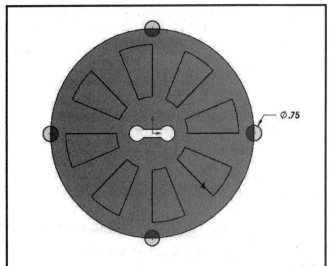

Figure 1-31. Sketching the Circles for the Guide Holes.

Go to the Features tab and select the Extruded Cut icon. For Direction1, just select **Through All,** and click the green (√) button to finish the part. Now view your model in an **Isometric** view, as shown in **Figure 1-32**.

If you would like to change the color of your model, click on the model name in the Feature manager tree and then select the colored ball in the menu bar. You can then assign any color you wish to the model.

You should now save your model. Pull down **File,** select your file folder, select **Save As,** type in the part name **COVER PLATE.sldprt**, then click **Save**. Insert the rendered Cover Plate image into your **Title Block** drawing sheet that was created in the introduction to Chapter 1 (See **Figure 1-32**. Follow the instruction given on **Page 1-7**. Now **SAVE** your drawing sheet to your directory as **COVER PLATE.slddrw**. Take note that the title of the part and the drawing are the same but the extension has changed. The solid model and the drawing are linked, meaning that any changes made to the solid model will automatically be updated on the drawing.

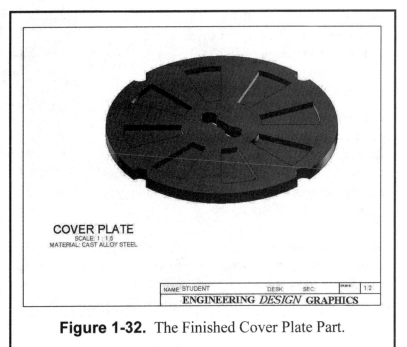

Figure 1-32. The Finished Cover Plate Part.

Exercise 1.3: WALL BRACKET

The sketch geometry used in Exercises 1.1 and 1.2, like line, circle, rectangle, and arc, are sufficient for designing many different kinds of parts. Sometimes more advanced 2D geometry, like an irregular curve, may be needed to complete a design. In the case of the Wall Bracket for this Exercise 1.3, you will learn how to use a spline.

Go to your folder and **Open** the file **ANSI-INCHES.prtdot**. In order to avoid corrupting the **ANSI-INCHES.prtdot** template go to **File - Save as:** under **File Name**, type **WALL BRACKET**, and under **Save as type**, Select **.sldprt** and select **SAVE**.

Click the **Front** plane in the "Feature Manager." Also click **Front** view in the Standard View toolbar to see the front plane in *blue* on the screen. Now click the **Sketch** icon in the upper left area of the screen. First sketch the straight line segments of the Wall Bracket outline using the **Line** tool. Refer to **Figure 1-33** for the proper **Dimension** for each line segment. Units are in inches and the upper left corner is at the origin.

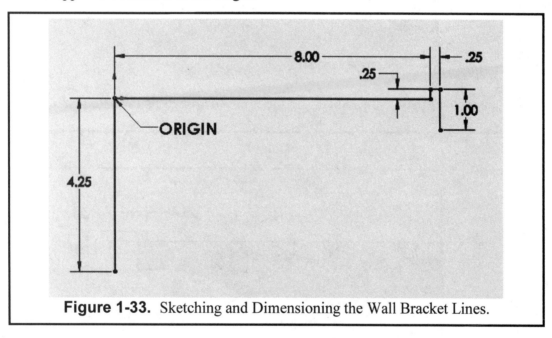

Figure 1-33. Sketching and Dimensioning the Wall Bracket Lines.

Now you will draw a spline from the bottom of the 4.25 inch line to the bottom of the 1.00 inch line. Click the **Spline** sketch icon (it looks like a sine wave). Click the **LMB** at the first bottom point. Then with the mouse pressed down, drag the spline over to the first intermediate point about one third of the way up and about one third of the way to the right. Release the **LMB** there, then press it down again and drag the spline to the second intermediate point about two thirds of the way up and over. Again release the **LMB** there, then press it down again and drag the spline to the final point, where you end the spline by double clicking the **LMB**. Refer to **Figure 1-34** for these four points' positions.

The two intermediate points are used to control the shape of the spline. With the **Select** cursor, pick the spline (it should highlight *blue* with its four points identified). Now pick intermediate point 1 and drag it out and down a little to create a slight bulge. Next, pick intermediate point 2 and pull it up a little to create a slight inflection (see **Figure 1-35**). *Note*: There are *no exact* X-Y coordinates for you to drag these intermediate points to. Just try it a few times until you achieve a shape that pleases you (for example, the shape in **Figure 1-35** is acceptable).

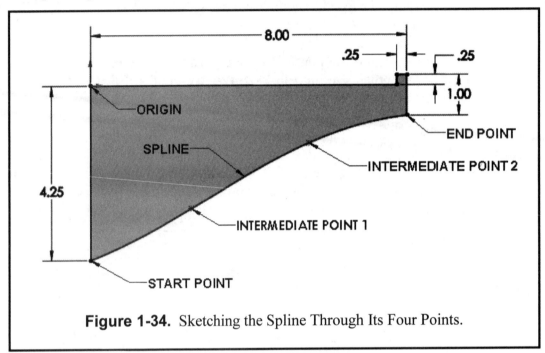

Figure 1-34. Sketching the Spline Through Its Four Points.

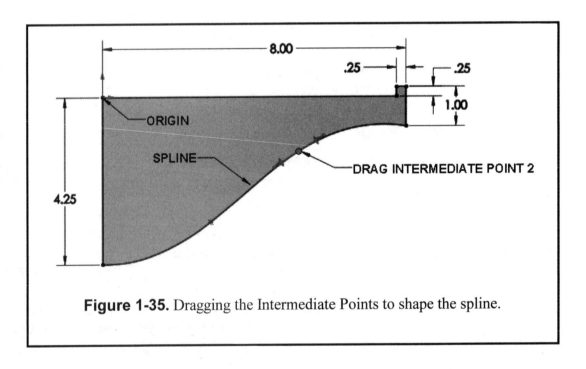

Figure 1-35. Dragging the Intermediate Points to shape the spline.

Now you will complete the left end of the Wall Bracket by adding the wall hooks. On the left edge, add the two wall hooks using the **Line** tool and the **Dimension** values given in **Figure 1-36**. You will have to **Trim** two places on the left edge to make the wall hooks contiguous with the rest of the profile. *Note: The bottom wall hook has the same dimensions as the top one.*

You will finish the Wall Bracket by adding a fillet to the sharp bottom corner on the right end. Since it was earlier associated with the *1.00 inch* dimension, **SOLIDWORKS** will give you a warning if you try to fillet it. So it is best to just first **Delete** that dimension. Now click the **Fillet** sketch icon. Key in a radius of **0.25** inches (refer back to **Figure 1-16**). Next, pick the vertical 1.00 inch right edge of the bracket and pick the spline close to its end point, and then click the green (√) check to execute the fillet command. The 2D profile is now complete and ready to be extruded. Activate the

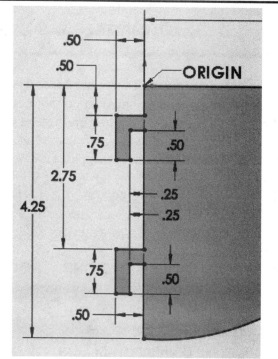

Figure 1-36. The Dimensions Used to Create the Wall Hooks on the Left Side.

Features Tab and **Select - Extruded Boss/Base**. Key in a **Blind** distance of **0.125** and click the green (√) check to complete the Wall Bracket, as shown in **Figure 1-37** in a **Trimetric** view.

If you would like to change the color of your model, click on the model name in the Feature manager tree and then select the colored ball in the menu bar. You can then assign any color you wish to the model.

Now **Save** the part as **WALL BRACKET.sldprt**. Insert the Wall Bracket onto a Title Block drawing sheet. Save as **WALL BRACKET.slddrw** and **Print** a hard copy for your instructor. See Page 1-7 for instructions.

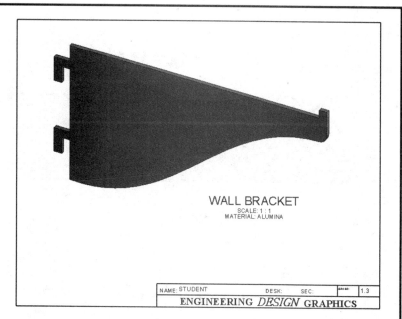

Figure 1-37. The Completed Wall Bracket in a Trimetric View.

Exercise 1.4: MACHINE HANDLE

In the previous exercises, the parts were designed using English units (inches). Metric, or System International (SI), units are also common in engineering practice. In Exercise 1.4, the Machine Handle will be designed using millimeters as the basic units (see **Figure 1-38**). In addition, because of the round symmetry of the machine handle, you will create a **Base Revolve** using the 2-D sketch rather than a base extrude.

Go to your folder and **Open** the file **ANSI-METRIC.prtdot**. In order to avoid corrupting the **ANSI- METRIC.prtdot** template go to **File - Save as:** under **File Name**, type **MACHINE HANDLE**, and under **Save as type**, Select **.sldprt** and select **SAVE**.

Next, Go to **TOOLS - OPTIONS – DOCUMENT PROPERTIES** and Select "**Grid/Snap.**" Make the following settings on this menu:

"Major grid spacing" = **20 mm**
"Minor lines per major" = **4**

Check the **Display Grid** box (√).

Also go to **System Snaps** and make sure "**Grid**" and the "**Snap**" functions are checked (√) on, then click the **OK** button.

In the **Feature Manager Tree**, select the **Front** plane and click the **Sketch Tab** and select the **Sketch** icon to start your sketch. The screen area should show a metric grid with spacing every 5 millimeters. You may need to **Zoom** in. Using the **Centerline, Line**, and **Centerpoint Arc** tools, sketch the enclosed 2-D profile shown in **Figure 1-38**.

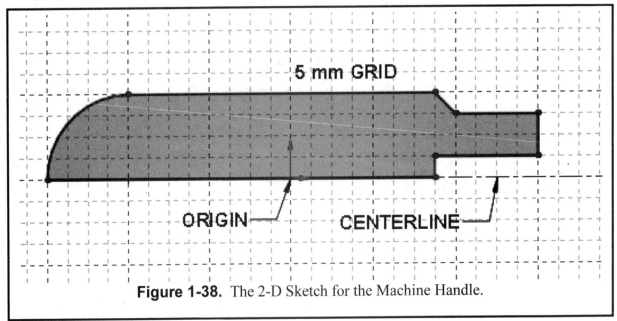

Figure 1-38. The 2-D Sketch for the Machine Handle.

Now you will revolve this simple sketch into a solid model. Activate the **Features** tab and **Select** the **Revolved Base/Boss** icon. The "Base Revolve" menu appears in the Feature Manager area as shown in **Figure 1-39**. The main choice you have here is the degrees of revolution about the centerline. Set it to a full revolution of **360°** and then click the green (√) check to close it. Your part should now appear as shown in **Figure 1-40** with an **Isometric** view.

There are many different designs for a Machine Handle, and the one here is quite typical. After studying it, however, you notice that the sharp outer edge on the right side poses a concern. SOLIDWORKS allows you to easily fix these design changes after the solid model has been built. One way is to simply go back and edit the original sketch.

Notice in the Feature Manager that SOLIDWORKS keeps a running tabulation of the features you used to build the part. One of them should read "**Revolve**," which is the operation you just completed. Left of this "**Revolve**" is a right facing arrow. Click on this arrow and down comes the name of "**Sketch1**." Right mouse click on this Sketch1 label and a short pull-down menu appears as shown in **Figure 1-41**. Left mouse click on the **Edit Sketch** option and the sketch appears again, with grids, on the screen. Select the **Front** view orientation to better see the sketch.

There are different ways to eliminate the sharp edge. One way is to use a chamfer; however, you probably cannot find a chamfer icon on the sketch editing toolbars. The chamfer icon is nested with the fillet icon. A **down arrow** next to the **Fillet** icon is another pull down menu and under that menu is the **Chamfer** icon. The "Sketch Chamfer" menu now appears. Set the "Chamfer parameters" as shown in **Figure 1-42**, including the chamfer distance of **3 mm**. Now click on the two lines that form the sharp edge corner, and then click the green (√) check to complete this chamfer operation.

The new feature is created, but the new model is not displayed yet. You need to rebuild the model first. Pull down **Edit**, and select the **Rebuild** option (it may be in the *customize menu* section of the pull-down menu).

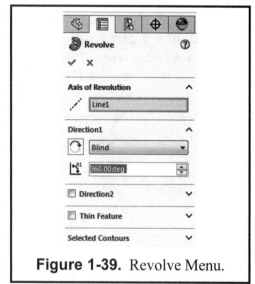

Figure 1-39. Revolve Menu.

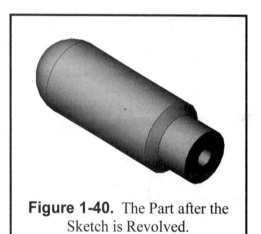

Figure 1-40. The Part after the Sketch is Revolved.

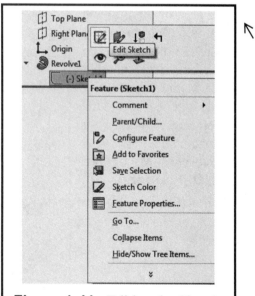

Figure 1-41. Editing the Sketch.

Note: There is also a small icon for the **rebuild command**. **It appears on the top toolbar and looks like a green and red traffic light.** The model is rebuilt now with a chamfer, as shown in the final picture (**Figure 1-43**) in an **Isometric** view with the **Shadows in Shaded** icon clicked **on**. Now you see how easy it is to make design changes to a solid model that is already built. You just simply edit the sketch that created the model in the first place.

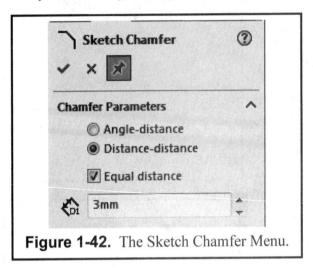

Figure 1-42. The Sketch Chamfer Menu.

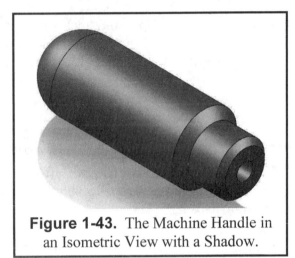

Figure 1-43. The Machine Handle in an Isometric View with a Shadow.

If you would like to change the color of your model, click on the model name in the Feature manager tree and then select the colored ball in the menu bar. You can then assign any color you wish to the model.

You should now save your model. Pull down **File,** select **Save As,** type in the part name **MACHINE HANDLE.sldprt**, and then click **Save**. Insert the rendered Machine Handle image into your **Title Block** drawing sheet, as shown in **Figure 1-44**. Add the title of the part in the open area of the drawing with a 20pt or 6mm letter height. Also add the Exercise number (**1.4**) in the upper right-hand box of the Title Sheet. Now **SAVE** your drawing sheet to your directory as **MACHINE HANDLE.slddrw**.

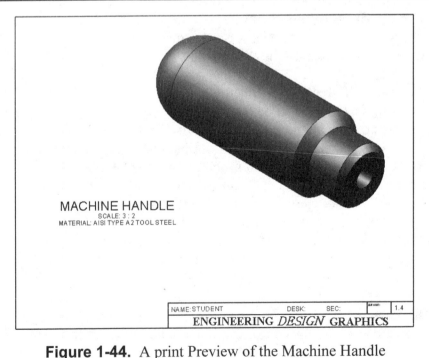

MACHINE HANDLE
SCALE: 3 : 2
MATERIAL: AISI TYPE A2 TOOL STEEL

NAME: STUDENT	DESK:	SEC:		1.4
ENGINEERING *DESIGN* GRAPHICS				

Figure 1-44. A print Preview of the Machine Handle on the Title Block Drawing Sheet.

Supplementary Exercise 1-5: SLOTTED BASE

Make a full-size sketch of the figure below in the **Top Plane** and extrude it for **0.50** inches. Insert it on a Title Block and title it **"SLOTTED BASE."**

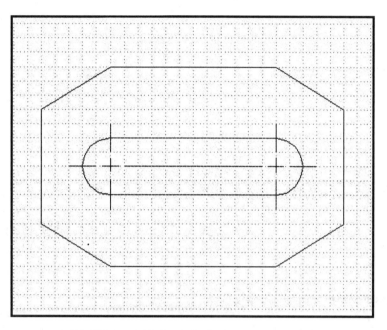

ASSUME THE GRID DIVISIONS TO BE 0.25 INCHES.

Supplementary Exercise 1-6: TRANSITION LINK

Make a full size sketch of the figure below in the **Front Plane**. Extrude it **0.375"**. Insert it onto a Title Block and title it **"TRANSITION LINK."**

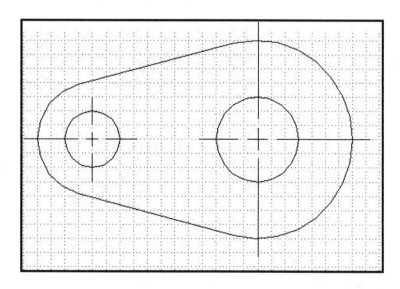

ASSUME THE GRID DIVISIONS TO BE 0.25 INCHES.

Supplementary Exercise 1-7: TEE BRACKET

Make a full-size sketch of the figure below in the **Top Plane** and extrude it to the height indicated. Insert it on a Title Block and title it **"TEE BRACKET."**

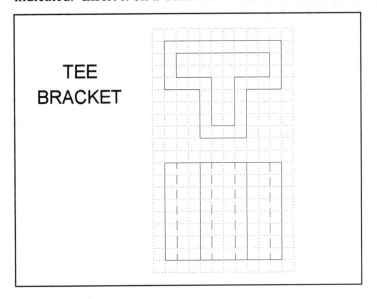

TEE
BRACKET

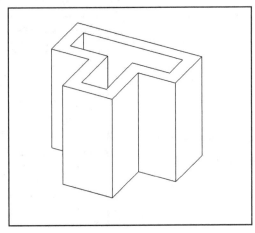

PICTORIAL OF THE TEE BRACKET

ASSUME THE GRID DIVISIONS TO BE 5 MM.

Supplementary Exercise 1-8: CABLE SPOOL

Make a full-size sketch of the **PROFILE** below in the **Front Plane** and do a **BASE - REVOLVE**. Insert it on a Title Block and title it **"CABLE SPOOL."**

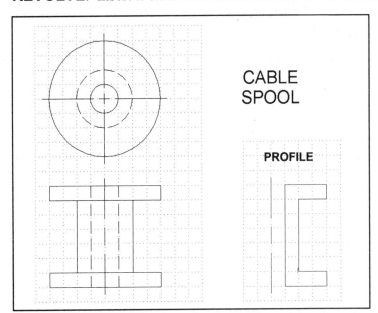

CABLE
SPOOL

PROFILE

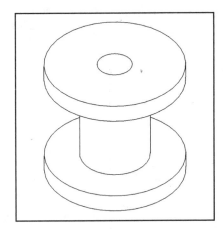

PICTORIAL OF THE CABLE SPOOL

ASSUME THE GRID DIVISIONS TO BE 1.00 INCH.

Computer Graphics Lab 2: 2-D Computer Sketching II

ADVANCED 2-D SKETCHING

In the first Computer Graphics Lab, you used some of the basic 2-D sketching capabilities of SOLIDWORKS. These first exercises concentrated on using items that were available on the sketching toolbars. You learned how to draw a Line, Circle, Rectangle, Arc, Polygon, Centerline, and Spline. You also learned how to edit the 2-D sketch using Dimensions, Trim, Mirror, Fillet, and Chamfer functions. In this Computer Graphics Lab 2, you will learn some more advanced 2-D sketching and editing features that are available in the vast SOLIDWORKS menu structure.

SKETCH ENTITY MENU

The sketch entities shown under the sketch tab are not the only ones available. Many of the icons have a small down arrow next to them. Each of these icons have additional options available for your use. These entities are also accessible under the **Tools – Sketch Entities** and are shown in **Figure 2-1**. Here you can find the following 2-D sketch entities:

> **Line**
> **Rectangles (several options)**
> **Parallelogram**
> **Slots (several options)**
> **Polygon**
> **Circle**
> **Perimeter Circle**
> **Centerpoint Arc**
> **Tangent Arc**
> **3 Point Arc**
> **Ellipse (several options)**
> **Partial Ellipse**
> **Parabola**
> **Spline**
> **Spline on Surface**
> **Point**
> **Centerline**
> **Text**

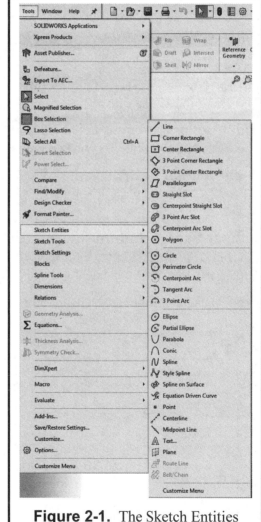

Figure 2-1. The Sketch Entities Menu.

Some of these 2-D entities are more common in engineering design than others, but hopefully you will have a chance to use each of them somewhere in one of your exercises.

SKETCH TOOLS MENU

All of the 2-D sketch editing functions are found under the **Sketch Tab**. On this menu you will find the following common editing functions:

Fillet is used to round a corner with a radius.

Chamfer is used to cut a corner at an angle.

Offset Entities is used to create another exact copy at a linear distance away.

Convert Entities converts an entity from an earlier feature to the current sketch.

Trim cuts away a piece of the entity.

Extend extends an entity to meet another entity.

Mirror copies a pattern around a centerline.

Dynamic Mirror first select the entity about which to mirror and then sketch the entities to mirror.

Jog Line moves a piece of the line up or down in a rectangular shape.

Construction Geometry converts entities to construction geometry or the converse.

Linear Sketch Pattern creates a rectangular array (row X column) of identical entities (see **Figure 2-3**).

Circular Sketch Pattern creates a radial (or polar) array of identical entities around a center point (see **Figure 2-4**).

Align is used to align a sketch and a grid point.

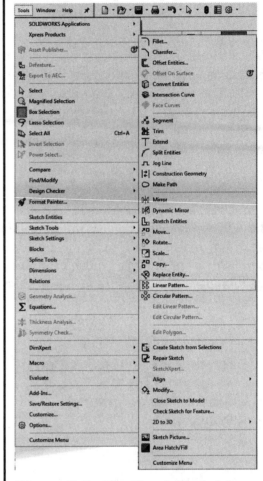

Figure 2-2. The Sketch Tools Menu.

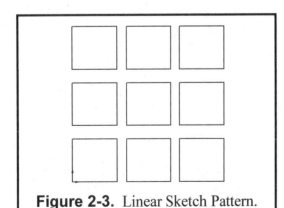

Figure 2-3. Linear Sketch Pattern.

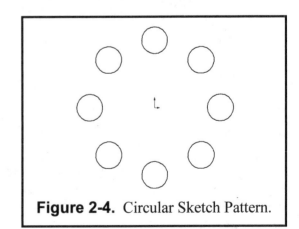

Figure 2-4. Circular Sketch Pattern.

Exercise 2.1: METAL GRATE

In Exercise 2.1, you will design a Metal Grate. The function of a metal grate is such that many identical slots are cut through it. Instead of drawing each slot separately, you will use an advanced sketching feature of SOLIDWORKS and create a rectangular array of these slots. Then you can simply extrude a base to create the beginning grate feature.

Start by going to your folder and **Open** the file **ANSI-METRIC.prtdot** because the dimensions of the Metal Grate are in Metric units. Immediately **SAVE AS – METAL GRATE.sldprt**. You will not need a grid for this exercise. Go to **Tools – Options - Document Properties** and click the **Grid/Snap** tab and make sure the "Display Grid" function is **not** checked (√) on, then click the **OK** button. Click the **Front** plane in the Feature Manager for the sketch plane.

Now activate the **Sketch Tab** and click the **Sketch** (pencil) icon to start your sketch. You will first draw two **Rectangle**s. The first one is the large outline of the grate and the second one is the initial small rectangular slot that eventually will be arrayed. Refer to **Figure 2-5** below for applying each **Dimension**. The overall size of the grate is **280** mm by **195** mm, and it is centered about the origin with its other two dimensions (**140** and **97.5**). The small slot is **20** mm by **35** mm and is **30** mm below the top and **30** mm to the right of the upper left corner. *Note:* After all the dimensions are applied, the lines turn black. This means that the geometry is completely fixed and constrained. Using the fillet command, add **3mm fillets** to the four corners of the small rectangle.

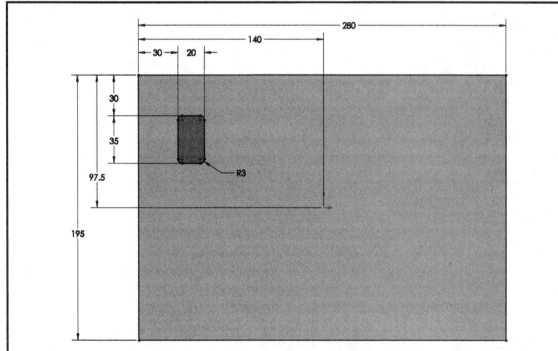

Figure 2-5. The Beginning Dimensions for the Metal Grate Centered at the Origin.

Now select the **Linear Sketch Pattern** icon in the sketch entities toolbar or pull down **Tools**; select **Sketch Tools** and then pick the **Linear Pattern** option. The "Linear Pattern Repeat" menu pops onto the screen. The Entities to Pattern box at the bottom of the menu is prompting you to select the lines and fillets of the small rectangle. The settings for this rectangular array operation are shown in **Figure 2-6** below. "Direction 1" is horizontal and will have **6** repeats. The horizontal spacing is **40** mm and the angle is **0** degrees. To activate "Direction 2" change the number of repeats to **3.** You will then be able to change the vertical spacing to **50** mm and the angle to **270** degrees. Notice that as you make adjustments to the linear table a **Preview** of the operation is shown before it is officially executed. If it is correct, click the **OK** button to complete the array. You should have a 6 x 3 array of slots that now can be extruded. You will have to click on the arrows to the left of the Y-axis button under Direction 2 to make the boxes drop below and onto the metal grate. Your pattern preview should look like the image in **Figure 2-7**.

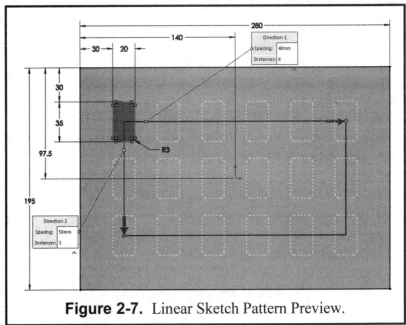

Figure 2-7. Linear Sketch Pattern Preview.

Figure 2-6. The Linear Sketch Step and Repeat Menu.

Select the **Features Tab** and select **Extruded Boss/Base**
Key in the following parameters:
      Type of Extrusion = **Blind**
      Distance 1 = **5** mm
Then click the green (√) check to close the menu. You will now have the base solid model of the grate, as shown in **Figure 2-8** in a **Trimetric** view.

The next step for the Metal Grate is to add a lip to the metal grate in order to provide a support when attached to the wall air duct. Click on this front surface of the Metal Grate. It should highlight *blue.* Then click on the **Sketch** tab and select the **Sketch Command** to add another sketch to the design. You have already drawn the outer rectangular profile, so you will borrow from it for the outer edge of the lip. Click the **Convert** sketch edit icon (it looks like a cube with a blue vertical edge). The outer lines now become part of your active sketch. Notice that they are all black lines since the geometry is already fixed.

Now click the top converted line (it turns *cyan*) and then click the **Offset Entities** icon (it looks like two bent parallel lines). Key in the offset value of **15 mm** and make sure the **Select Chain** box is checked (√). If the 15 mm offset is previewed on the outside, check (√) the **Reverse** box in the menu box so the offset is to the *inside* of the original lines and then click the green

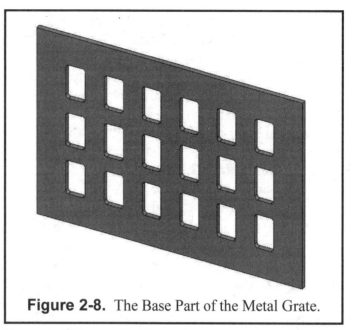

Figure 2-8. The Base Part of the Metal Grate.

(√) checkmark to create the offset, as shown in **Figure 2-10**. Also, notice that the offset command places a small 15 mm dimension on your sketch to indicate the offset value. You could now simply click on that dimension directly, key in a new dimension value, and instantly change the offset to a new value. But for now leave it at 15 mm. Fillet the inside corners of the offset pattern to **5mm** as shown in **Figure 2-10**.

Before you perform the Extrude command you may want to go to an Isometric view in order to see which direction you are extruding. Select the **Features** icon and select **Extrude.** When the **Extrude** menu appears, key in the following parameters:

Type of Extrusion = **Blind**

Distance 1 = **5 mm**

Click the green (√) check to complete the boss.

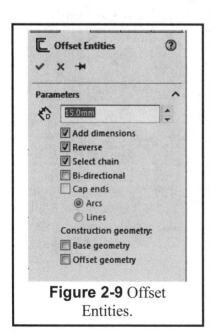

Figure 2-9 Offset Entities.

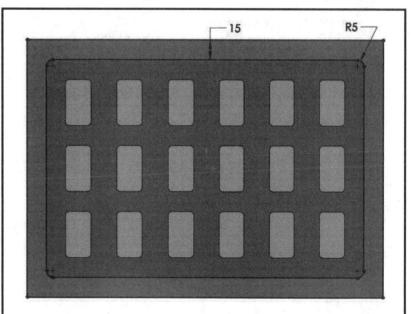

Figure 2-10. Converting the Front Edges, Offsetting them by 15 mm and Filleting the Inside Corners by 5mm.

Now you need to add four attachment holes to the corners of the grate. **Select** the raised rim (it will turn *blue*). Click on the **Sketch Tab** and activate the **Sketch icon** and draw a **Circle** in the upper left corner. Use the **Dimension** values supplied in **Figure 2-11** for the circle diameter (**8** mm) and position from the corner (**9**mm x **9**mm).

Now draw three more **Circle**s in the other three corners. **Dimension** them to have the same diameter (**8**) and same relative position (**9** x **9**) from each corner. *Or*, now that you are an expert with a rectangular array, use the **Linear Sketch Pattern** operation instead. If you use this function, then the horizontal distance of the **2** items is **262** mm and the angle is **0** degrees. The vertical distance of the **2** items is **177**mm and the angle is **270** degrees. *Either way*, when you are finished you should have circles at the four corners and click the green (√) check to execute the Linear array.

Change your viewpoint to a Trimetric view. Now activate the **Features** icon and select **Extruded Cut**. Select the extrude type to be **Through all** and click the green (√) check to execute the cut. The four corner attachment holes are now created on the grate.

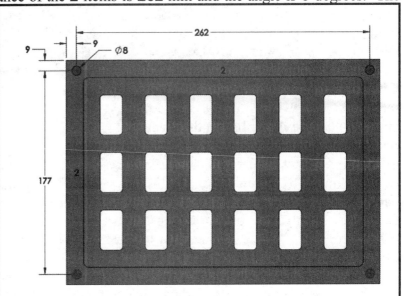

Figure 2-11. The Dimension Values for the Small Holes.

The part is now complete and you can view the lip feature more clearly by using the **Rotate View** icon as shown in **Figure 2-12**.

If you would like to change the color of your model, click on the model name in the Feature manager tree and then select the colored ball in the menu bar. You can then assign any color you wish to the model.

Return to a **Trimetric** View of your part as shown in **Figure 2-12**. You should now save your model. Pull down **File**, select **Save As,**

Figure 2-12. The Final Design of the Metal Grate in a Trimetric View.

type in the part name **METAL GRATE.sldprt**, and then click **Save**. Open your copy of **TITLE BLOCK – METRIC.drwdot** and immediately **SAVE AS – METAL GRATE.slddrw**. Now insert the rendered Metal Grate image into your **Title Block** drawing sheet that was created in Chapter 1 and **Print** it on this sheet (see **Figure 2-13**).

Print a hard copy to submit to your lab instructor.

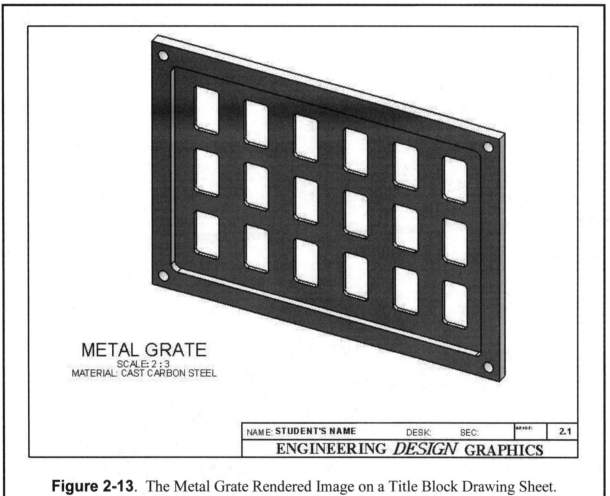

METAL GRATE
SCALE: 2 : 3
MATERIAL: CAST CARBON STEEL

| NAME: STUDENT'S NAME | DESK: | SEC: | GRADE: | 2.1 |

ENGINEERING *DESIGN* GRAPHICS

Figure 2-13. The Metal Grate Rendered Image on a Title Block Drawing Sheet.

Exercise 2.2: TORQUE SENSOR

In Exercise 2.2 you will design a Torque Sensor casing. Since it is a circularly symmetrical object, you will employ some of the advanced editing features like circular array. Go to your folder and **Open** the file **ANSI-INCHES.prtdot**, and immediately **SAVE AS – TORQUE SENSOR.sldprt**. Select the **Tools**, **Options**, **Document Properties** menus and Select "**Grid/Snap**." Make the following settings on this menu: "**Major grid spacing**" = **1.00**, "**Minor lines per major**" = **4**. Also go to System Snaps and make sure "**Display Grid**" and the "**Snap**" functions are checked (√) on, then click the **OK** button. Make sure the **Units** are in **Inches**. Then click **OK.**

The circular features of the Torque Sensor are on the top and bottom surfaces. But the main body is also round and can be created by a 360 degrees revolution of a profile that has been drawn on a frontal plane. So click on the **Front** plane in the Feature Manager tree. Click on the **Sketch Tab** and select the **Sketch Icon** and the sketching grid appears with minor grids spaced every 0.25 inches. Also make sure you are viewing this from the **Front** view orientation.

First draw a **Centerline** vertically through the origin. Next, use the **Line** tool to sketch the completely enclosed profile that is depicted in **Figure 2-14**. This design will yield a part that is 2.50 inches tall and 4.00 inches in diameter on the top and bottom surfaces.

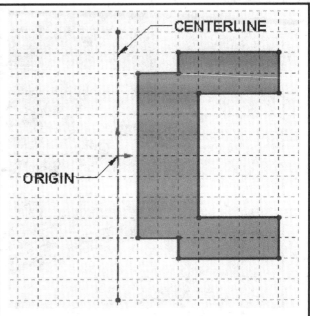

Figure 2-14. The Initial Lines to Revolve for the Base Part.

Go to the **Features** tab and **Select** the **Revolved Boss/Base** icon. Make sure that the centerline is selected for the Axis of Revolution. Key in the full revolution value of **360°** and click the green (√) check to perform the revolution. The circular base part appears as shown in **Figure 2-15** in an **Isometric** view.

The next design step is to create a circular array of eight holes around a bolt circle on the top surface of the part.

Figure 2-15. The Base Part after the Revolution.

Click on the *top* **surface of the part and it should highlight** *blue*. Also select a **Top** view orientation. Then select the **Sketch** icon. **Draw a Circle** that is **3.25** inches in diameter, and make sure you select **"for construction"** in the feature manager tree. Then draw a horizontal center line from the origin and to the right. The intersection of these two entities defines the center of the first of eight holes, thus resulting in a radius of 1.625. Or you can go to the **Document Properties** menu and on the **Grid/Snap** tab change the "Minor lines per major" value to **8**, thus resulting in a one-eighth inch grid. Also on the **Units** tab

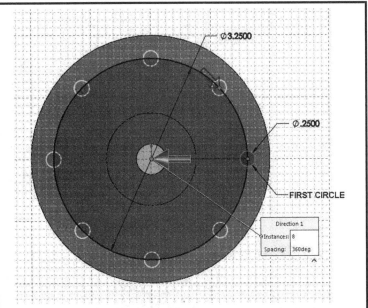

Figure 2-16. Sketching the First Circle and Executing a Circular Array of Eight Holes.

change the decimal places to **3**. Click **OK** and the grid should now be updated to the new values. Now locate the center of the first **Circle** on the grid and draw it with a diameter of **0.25**. Use **Figure 2-16** to aid you.

Select the circle (it should highlight *blue*). Click on the **down arrow** next to **Linear Sketch Pattern** to select the **Circular Sketch Pattern** option. The "Circular Pattern" menu appears on the screen. Referring to **Figure 2-17**, set the parameters for this circular array. The "Radius" is **1.625** from the center (**0,0**). The "Step Number" is **8** for a "Total angle" of **360°**. The spacing is "Equal" checked (√) on. Click **Preview** to see if everything is correct, and then click the **OK** button. You now have a bolt circle of 8 holes as previewed earlier in **Figure 2-16**. You are now ready to cut these holes through the entire base part.

Switch to the **Shaded** model mode and to an **Isometric** view to better see the next operation. Select the **Features** tab and select **Extruded Cut**. On the "Cut Extrude" menu select **Through all** for the direction and click the green (√) check to execute the cut extrusion all the way through the model. Use the **Rotate View** icon to see that the holes are indeed

Figure 2-17. The "Circular Pattern" Menu to Create the Bolt Circle Holes.

all the way through the bottom of the model. If so, then the model is complete as shown in **Figure 2-18**.

If you would like to change the color of your model, click on the model name in the Feature manager tree and then select the colored ball in the menu bar. You can then assign any color you wish to the model.

You should now save your model. Pull down **File,** select **Save As,** type in the part name **TORQUE-SENSOR.sldprt**, and then click **Save**. Open your **TITLE BLOCK – INCHES.drwdot** and **SAVE AS: TORQUE SENSOR.slddrw.** Now insert the rendered Torque Sensor

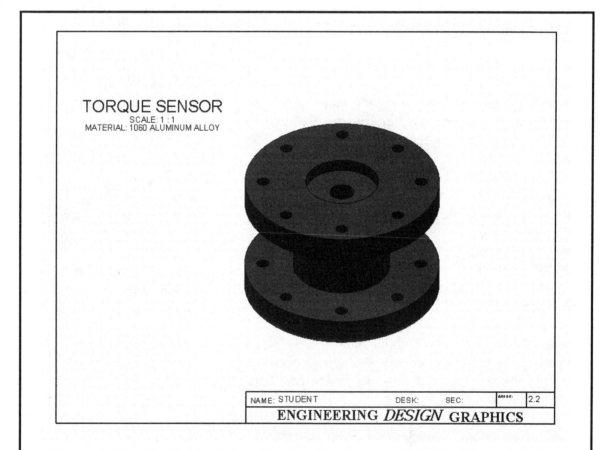

Figure 2-18. The Finished Model of the Torque Sensor in an Isometric View.

image into your **Title Block** drawing sheet that was created in Chapter 1 and **Print** it on this sheet (see **Figure 2-19**).

Figure 2-19. The Torque Sensor Rendered Image on a Title Block Drawing Sheet.

Exercise 2.3: SCALLOPED KNOB

In Exercise 2.3, you will design a Scalloped Knob that has some complicated geometry around its edges. This particular knob design will be a hexagon type. Since the hexagonal features are equally spaced around the center of the knob, you can use a circular array function.

Start by going to your folder and **Open** the file **ANSI-INCHES.prtdot** and immediately **SAVE AS – SCALLOPED KNOB.sldprt**. Go to **TOOLS – OPTIONS – DOCUMENT PROPERTIES** and change the **UNITS** to three decimals. Select the **Front** plane for the sketch. Then start a new **Sketch**. Complete the initial geometry of the sketch according to **Figure 2-20**. Using the **Line** tool, draw two vertical lines and cap them off with a horizontal line that touches their top ends. **Fillet** the top two corners with a **0.10** radius. Use the **Dimension** tool to completely fix the geometry by applying the dimensions shown in **Figure 2-20**. Include dimensions that relate to the origin. When the geometry is fixed, all lines turn *black*.

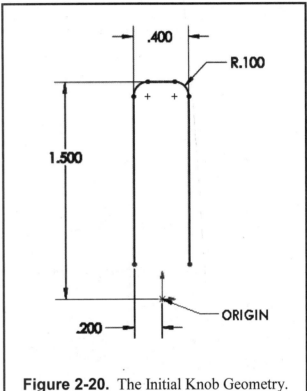

Figure 2-20. The Initial Knob Geometry.

Now array this pattern in a circle to form a hexagonal layout. There is a **pull-down arrow** next to the **Linear Sketch Pattern**. When you select it you will see the **Circular Sketch Pattern** option. The "Circular Pattern" menu appears on the screen. Set the "Step Number" to **6** for a "Total angle" of **360°**. The spacing is "Equal" checked (√) on. Activate the "Entities to Pattern" box and **select the three straight lines and the two fillets**. Click **Preview** to see if everything is correct, and then click **OK**. You now have an array that is the beginning of the sketch for the knob outline. Notice that some of the lines may overlap as can be seen in **Figure 2-21**. You may want to trim the intersecting lines; however, that is not necessary to complete the remainder of the exercise.

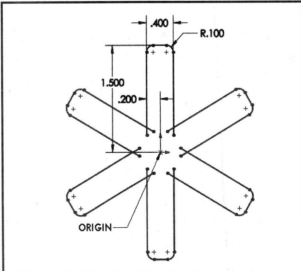

Figure 2-21. The Sketch after Arraying the Pattern.

Next, you will fillet the six sharp inner corners to create the scallop effect. Pick the **Fillet** sketch icon and key in a fillet radius of **0.45** in the "Sketch Fillet" parameter box. Now pick two intersecting lines. A large 0.45 radius is made and a small dimension is attached to show the fillet value. Repeat this filleting process on the remaining five sharp inner corners. When you are finished, your sketch should look like **Figure 2-22**.

Select the **Features** tab and select **Extruded Boss/Base**. Extrude the sketch to a **Blind** depth of **0.375** inches. Click the green check (√) to close the operation. When finished, view the part in a **Trimetric** orientation as shown in **Figure 2-23**.

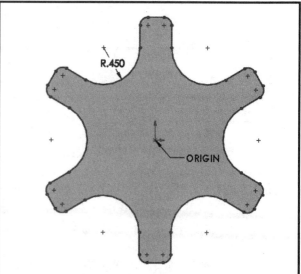

Figure 2-22. The Finished Sketch after Filleting Six Sharp Inner Corners.

You now can finish the part by adding the attachment base. **Click the front surface to highlight it in *blue*.** Set your view orientation to **Front**. Click the **Sketch** icon and draw a **Circle**, centered at the origin. **Dimension** the circle to be **1.125** inch in diameter. Now draw a **Hexagon** at the origin. Check **Inscribed Circle** and set the diameter to **0.625**. Select the **Features** icon and select **Extrude**. It is advisable to **go to an Isometric view** when executing an extrusion of any kind. Extrude the sketch to a **Blind** depth of **.75** inches away from the front surface.

Select the **Dimetric** view to see the inside of the hexagonal hole. **Select** the visible surface of the knob and with the **Features - Fillet** enter **0.05"** to remove the sharp edges of the knob. Repeat the process for the back surface of the knob.

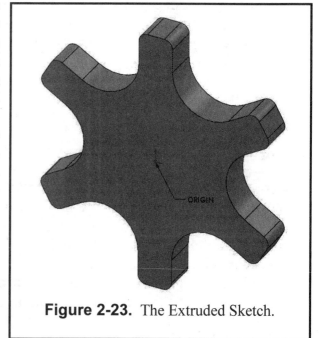

Figure 2-23. The Extruded Sketch.

If you would like to change the color of your model, click on the model name in the Feature manager tree and then select the colored ball in the menu bar. You can then assign any color you wish to the model.

Now save your model to your designated folder. Pull down **File**, select **Save As**, type in the part name **SCALLOPED KNOB.sldprt**, and then click **Save**. Open your **TITLE BLOCK – INCHES.drwdot** and immediately **SAVE AS – SCALLOPED KNOB.slddrw**. Now insert the rendered Scalloped Knob image onto your **Title Block** drawing sheet that was created in Chapter 1 and **Print** it on this sheet (see **Figure 2-26**).

Print a hard copy to submit to your lab instructor.

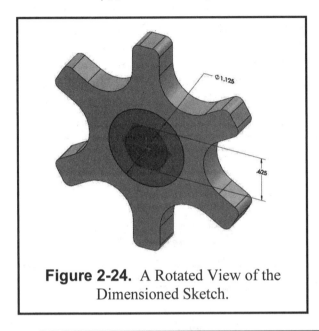

Figure 2-24. A Rotated View of the Dimensioned Sketch.

Figure 2-25. The Shaded Model of the Scalloped Knob.

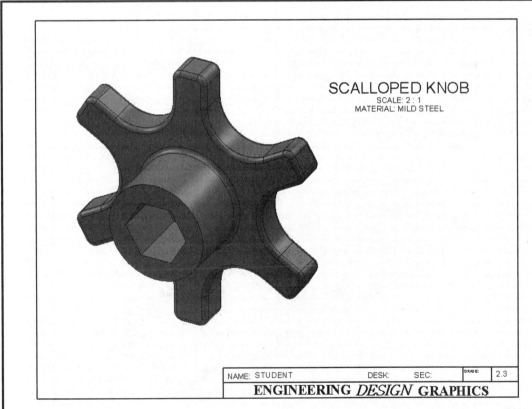

SCALLOPED KNOB
SCALE: 2 : 1
MATERIAL: MILD STEEL

NAME: STUDENT		DESK:	SEC:	GRADE:	2.3

ENGINEERING *DESIGN* GRAPHICS

Figure 2-26. The Scalloped Knob Rendered Image on a Title Block Drawing Sheet.

Exercise 2.4: LINEAR STEP PLATE

In Exercise 2.4, you will design a Linear Step Plate used for linear motion control in machinery. There are a lot of holes on this plate, and you will find the linear array and mirror functions to be quite helpful. Start by going to your folder and **Open** the file **ANSI-INCHES.prtdot**; immediately **SAVE AS – LINEAR STEP PLATE.sldprt**. Select the **Right Plane** as the drawing plane and the **Right** view orientation to see it head on. Then start a new **Sketch**. Draw a vertical centerline through the origin. Go to **TOOLS – Sketch Tools**, and Select **Dynamic Mirror**. Now sketch the right half of the profile shown in Figure 2-27. Each line drawn on the right side of the centerline will be duplicated on the left. Use the **Dimension** tool to set the geometry by applying the dimensions shown in the **Figure 2-27**, including the dimension to the origin.

Select the **Features** icon and select **Extruded Boss/Base**. On the "Base Extrude" menu, set the extrude parameters as shown in **Figure 2-28**:

Direction 1: **Blind, 4.2000 in.**
Direction 2: **Blind, 4.2000 in.**

OR

Extrude the Sketch 8.4 in. MID-PLANE

Notice that you can preview this operation in an **Isometric** view on the screen. Then click the green (√) check to close the menu and execute the extrusion in two directions. The base part looks like **Figure 2-29**.

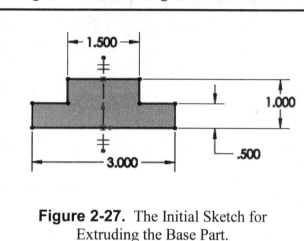

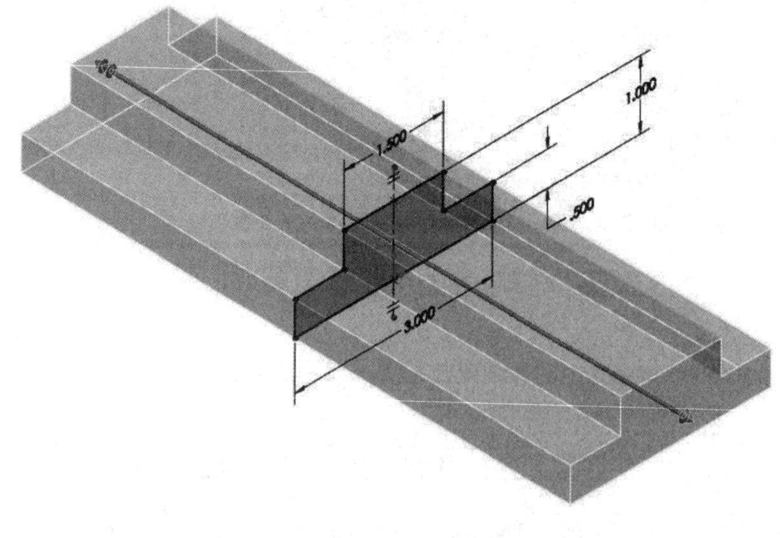

Figure 2-27. The Initial Sketch for Extruding the Base Part.

Figure 2-28. Extruding the Sketch in Both Directions.

Now you will create some linear holes. **Pick the top surface of the small step on the front side** (see **Figure 2-29**). **It should highlight *blue***. In a **Top** view, click **Sketch** and draw a small **Circle** on the surface as shown in **Figure 2-30**. Use the **Dimension** tool to add the three dimensions given to fix it:

Diameter = **0.300**
From center origin = **1.125**
From center origin = **3.000**

Now you will linearly repeat that circle. **Select** the circle (it should turn *cyan*). Select the **Linear Sketch Pattern** icon at the top of the screen. The "Linear Sketch Step and Repeat" menu pops onto the screen. Key in the following parameters:

Direction 1:
Number = **6**
Spacing = **1.2000**
Angle = repeat to right side

Direction 2:
Number = 1

You now should have six circles on the front step surface. You need to add six more circles to the back step surface. You can mirror them.

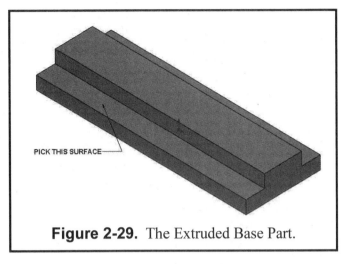

Figure 2-29. The Extruded Base Part.

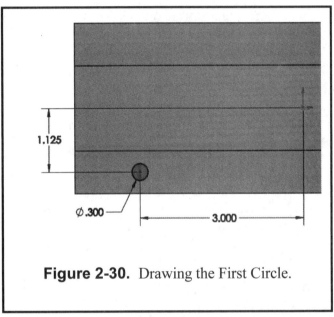

Figure 2-30. Drawing the First Circle.

Draw a horizontal **Centerline** across the origin (Note the "—" symbol on your cursor means horizontal). Click the **Mirror** sketch icon and the mirror menu pops onto the screen. For the "Entities to Mirror," select the six circles just created in the Linear pattern and in the "Mirror About" box

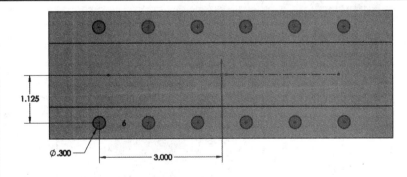

Figure 2-31. Linearly Repeating and Mirroring the Circles.

select the centerline drawn through the origin. The selected items will get mirrored about the centerline, as shown in **Figure 2-31**.

Select the **Features** icon and select **Extruded Cut**. Use the **Through All** option and click the green (√) check to close the menu. You have now drilled the small holes all the way through the plate's steps. You now need to bore some counterbore holes a quarter of the way down the small through holes. *Note:* This design feature is called a

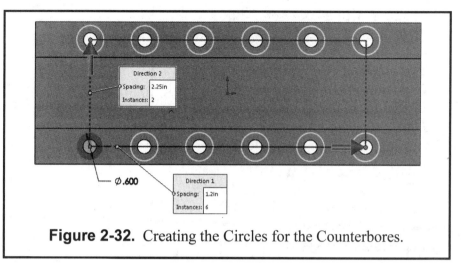

Figure 2-32. Creating the Circles for the Counterbores.

"Counterbore" and SOLIDWORKS has a special "Wizard" that can create it. However, we will leave that "Wizard" for a later lab.

Select the top surface of the front step again. **Sketch** a **Circle** on that surface. Then add a relation to make the circle concentric with the hole beneath it (the diameter is **.60**). Now repeat the exact same process as before to get the twelve circles for the counterbore holes.

- o **Select** the new circle.
- o Execute a **Linear Pattern** to get the front **6** circles at **1.20** inches apart in **Direction 1**.
- o **In Direction 2** increase instances to **2** at a distance of **2.25 inches**.

Select the **Features** icon and select **Extruded Cut**. Use the **Blind** option to a depth of **0.125** inches into the material. Click the green (√) check to close the menu. You now bored the counterbores into the plate's two steps, as shown in **Figure 2-33** in a **Rotated View**.

Note: Sometimes you might make a mistake with a **FEATURE** operation like this one. You can simply right mouse click on its name in the Feature Manager and select the **Edit Feature** option on the menu. See **Figure 2-34**.

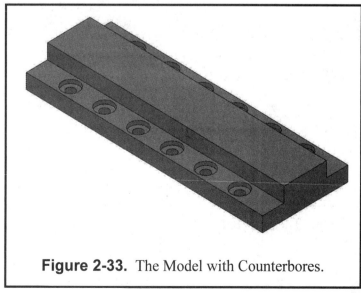

Figure 2-33. The Model with Counterbores.

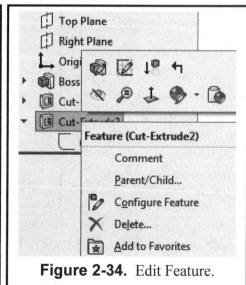

Figure 2-34. Edit Feature.

The next design requirement is to create four holes on the top of the plate. Pick a **Top** view and select the top surface to **Sketch** on. Draw a first **Circle** with the three **Dimension** values given in **Figure 2-35**. Use a **Linear Pattern** operation to get a second circle **1.2000** inches from the first circle. Draw a vertical **Centerline** through

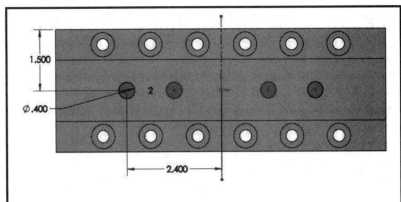

Figure 2-35. Creating the Holes in the Top Surface.

the origin (a | appears on the cursor). Then **Mirror** the two circles. This results in four circles as shown in **Figure 2-35**.

It is advisable to **go to an Isometric view** when executing an extrusion of any kind. Select the **Features** icon and select **Extruded Cut**. Use the **Through All** option and click the green (√) check to close the menu. You now drilled the small holes all the way through the thick part of the plate. **Select the top surface** again to begin a new sketch. You will now add two counterbore slots. Select the **Slot** icon in the sketch menu. Select the **straight slot** type. Select the center of the left circle on the top plane and the one immediately to its right. This will make the slot concentric with the two circles to the left of the center. You can use an identical process to sketch the slot to the right of the center. Dimension the arcs of both slots to have a **Radius of .40**. Then **Cut Extrude** them to a **Blind** depth of **0.2500**. These counterbore slots are shown in **Figure 2-37**. To finish the step plate, chamfer the three horizontal edges on both ends of the model. Activate the **Features tab** and under the pull down menu of the **Fillet, Select Chamfer**. Set the

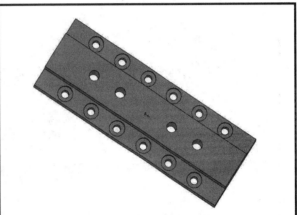

Figure 2-36. The Four Through Holes in the Top Surface.

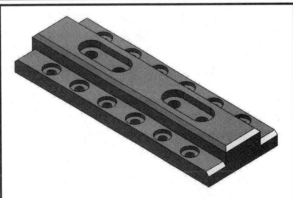

Figure 2-37. The Completed Linear Step Plate.

chamfer value to **.125**, then select the three top horizontal edges of the ends of the step plate and the two long edges of the top surface. Click the green (√) check to complete the exercise.

In the Feature Manager Tree, Right click on **Edit Material**, expand the **Copper Alloy** materials category and assign **Brass** to the Linear Step Plate.

The part is now finished. Return to an **Isometric** view of the finished part as shown in **Figure 2-37**. Pull down **File,** select **Save As,** type in the part name **LINEAR STEP PLATE.sldprt,** and then click **Save**. Open your **TITLE BLOCK – INCHES.drwdot** and immediately **SAVE AS – LINEAR STEP PLATE.slddrw**. Now insert the rendered Linear Step Plate image onto your **Title Block** drawing sheet that was created in Chapter 1 and **Print** it on this sheet (see **Figure 2-38**).

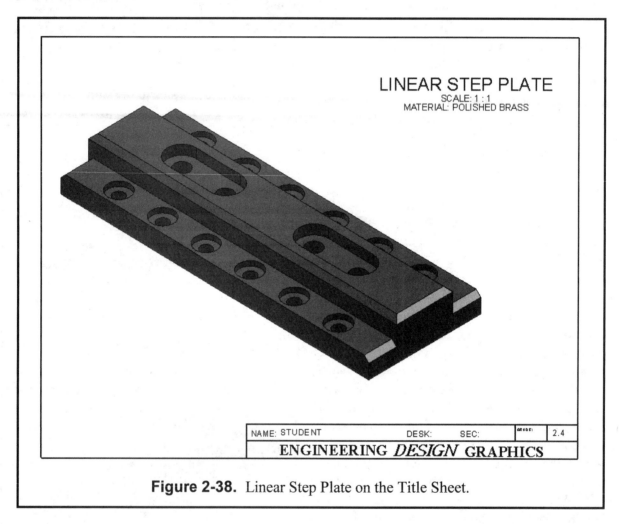

Figure 2-38. Linear Step Plate on the Title Sheet.

SUPPLEMENTARY EXERCISE 2-5: FLANGE

Using the **Revolve** command in the **Front Plane**, the **Circular Step and Repeat** commands learned in Unit 2, in the **Top Plane** Build the Flange and extrude it according to the grid divisions. Insert it on a Title Block and title it **"FLANGE."**

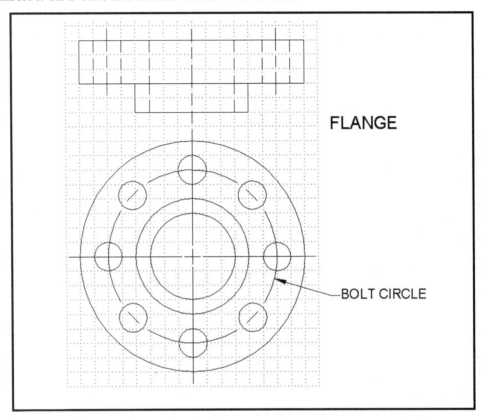

ASSUME THE GRID DIVISIONS TO BE 0.50 INCHES.

SUPPLEMENTARY EXERCISE 2-6:
STEEL VISE BASE

Make a full size model of the figure below using the commands learned in Unit 2. Insert it onto a Title Block and title it **STEEL VISE BASE**.

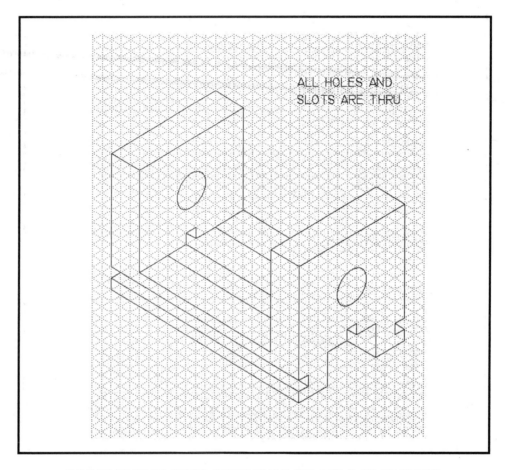

ALL HOLES AND
SLOTS ARE THRU

ASSUME THE GRID DIVISIONS TO BE 0.25 INCHES.

Computer Graphics Lab 3:
3-D Solid Modeling of Parts I

In your first two Computer Graphics Labs #1 and #2, you created some simple solid parts by using 2-D sketches. Almost all of the geometric size and shape data were defined in the 2-D sketch and you simply extruded or revolved the sketch to get the 3D solid model. While this is the normal beginning procedure, SOLIDWORKS contains a variety of advanced commands in which your geometric data can be created or edited directly in 3D space. In this Computer Graphics Lab #3, you will start to create solid models using both 2D sketches and 3D features.

ADDING SKETCH RELATIONS

The 2D sketch will continue to be the normal mode for initiating the construction of a solid model. You have already learned how to draw a variety of sketching entities, and how to edit and dimension them. One more capability in the 2D sketching mode is to **Add Relations**. The Relations toolbar is shown in **Figure 3-1** and can be placed on screen by going to **View – Toolbars** and selecting **Dimensions/Relations**. Relations fix the characteristics of one or more sketch entities. The following are the common relations you may find useful in making a 2D sketch.

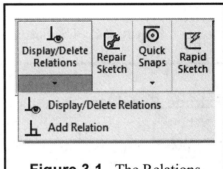

Figure 3-1. The Relations Toolbar.

Horizontal makes one or more lines horizontal.

Vertical makes one or more lines vertical.

Collinear makes two or more lines lie on the same infinite line.

Co radial makes two or more arcs share the same center point and radius.

Perpendicular makes two lines perpendicular to each other.

Parallel makes two or more lines parallel to each other.

Tangent makes an arc, ellipse, or spline, and a line or arc tangent to each other.

Concentric makes two or more arcs, or a point and an arc to share the same center point.

Midpoint makes a point on a line to remain at the midpoint of the line.

Intersection makes two lines and one point to remain at the intersection of the lines.

Coincident makes a point and a line, arc, or ellipse to lie on the line, arc, or ellipse.

Equal makes two or more lines or two or more arcs have the same lengths or radii.

Symmetric makes a centerline and two points, lines, arcs, or ellipses to be equidistant from the centerline.

Fix makes the entity's size and location fixed.

THE FEATURES TOOLBAR

The "Features" toolbar is located on the top of your screen and is shown in **Figure 3-2**. These represent the main tools available in SOLIDWORKS to create and edit parts in 3D space. Below are descriptions of these features. _Note:_ In SOLIDWORKS, the first solid feature that you build is called a base, and after that they are called bosses for that part.

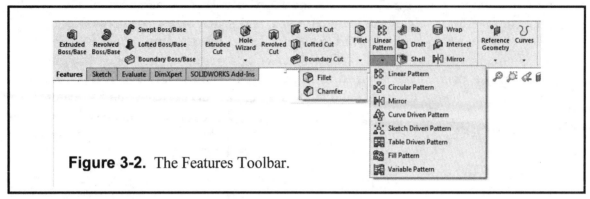

Figure 3-2. The Features Toolbar.

Extrude Boss/Base creates a base or boss by linearly extruding a sketch in an orthogonal direction.

Revolve Boss/Base creates a base or boss by revolving a sketch around a centerline.

Extruded Cut subtracts material from a solid body by linearly extruding a sketch through it.

Revolved Cut subtracts material from a solid body by revolving a sketch around a centerline.

Sweep creates a base, boss, cut, or surface by moving a profile along a designated path.

Loft creates a feature by making transitions between profiles.

Fillet creates a rounded internal or external face on the part by picking an edge.

Chamfer creates a beveled feature on selected edges or a vertex.

Rib adds material of a specified thickness determined by a contour and an existing part.

Shell hollows out the part, and leaves open the faces you select.

Draft tapers faces of a part using a specified angle.

Wrap wraps a closed sketch contour onto a face.

Dome adds a dome to a planer or non-planer face.

Hole Wizard allows you to quickly add different types of common holes to your part.

Linear Pattern creates multiple instances of selected features along one or two linear paths.

Circular Pattern creates multiple instances of one or more features uniformly around an axis.

Mirror Feature creates a copy of a feature (or multiple features) that is mirrored about a plane.

Reference Geometry creates reference geometry like planes, axis, coordinate system, and points.

Curves - creates different types of curves including spiral and helix.

Exercise 3.1: CLEVIS MOUNTING BRACKET

In Exercise 3.1, you will design a Clevis Mounting Bracket. You will start with a 2D sketch of the main outline in a frontal view. Because of its design nature, certain features of the Bracket must remain fixed while other features and dimensions can be varied to accommodate design changes. For example, a hole must remain concentric with an outer arc, but the height of the hole from the bottom base could vary. So you will need to add some relations to the sketch that will fix or "constrain" the geometry. In addition, in this exercise you will use some 3D editing functions, like "Mirror Feature," to complete the design.

Go to your folder and **Open** the file **ANSI-INCHES.prtdot**. In order to avoid corrupting the **ANSI-INCHES.prtdot template**, go to **File - Save as:** under **File Name**, type **CLEVIS MOUNTING BRACKET**, and under **Save as type**, select **.sldprt** and select **SAVE**. Click the **Right Plane** in the Feature Manager (it turns *blue*). The Clevis Bracket is symmetric about the right plane in 3D space. So the initial Sketch1 needs to be constructed on a plane that is parallel to the right plane but a distance from it. To better see this operation, click to **Isometric** view orientation. Pull down **Insert**, select **Reference Geometry,** and

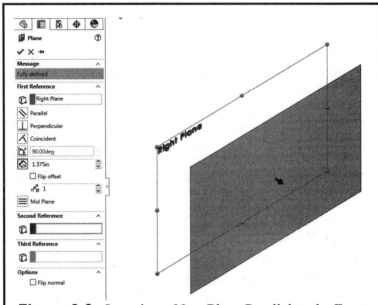

Figure 3-3. Inserting a New Plane Parallel to the Front Plane and 1.375 Inch in Front of It.

then select **Plane**. The "Plane" menu appears as shown in **Figure 3-3**. Indicate the distance, which should be **1.375** inch, then close the "Plane" menu with (√).

You now have a new plane in the Feature Manager tree called **Plane1**. Click on this new plane and then click the **Sketch Tab** and then the **Sketch icon** to start your 2-D sketch. Use the **Line**, **Circle**, and **Arc** sketch tools to construct the rough 2D profile shown in **Figure 3-4**. You do not have to be very accurate right now with your sketch because you will be defining relations and dimensions that will constrain the geometry. Just make sure the sketch is a little above the origin as shown in **Figure 3-4**.

Now click on the **Display/Delete Relations** pull down tab and select **Add Relations** icon on the sketch toolbar (it is the icon with the perpendicular symbol). Click the bottom base line and it turns *cyan*. Now also notice that the "Add Relations" menu appears in the Feature Manager area as shown in **Figure 3-5**. It has three menu boxes:

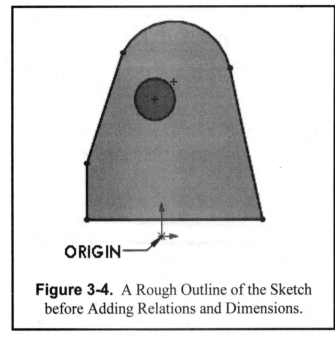

Figure 3-4. A Rough Outline of the Sketch before Adding Relations and Dimensions.

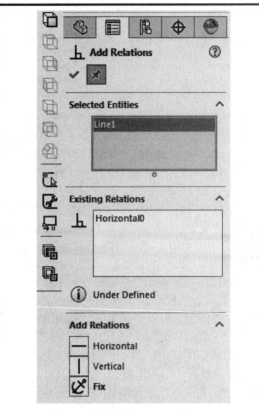

Figure 3-5. The Add Relations Menu.

"Selected Entities" displays the currently selected entity(ies);

"Existing Relations" displays any pre-existing relations before you clicked the item(s);

"Add Relations" lists all the possible relations the item(s) can have.

Add the **Horizontal** relation to this bottom line and click the green check mark (√) to close the menu. The bottom line is now horizontal. *Note:* Depending on how you drew the first line, it may already be horizontal because the smart cursor sometimes infers the intent of the designer.

Now study all the remaining relations to add to the sketch, as shown in **Figure 3-6**. Each time click the **Add Relations** icon, pick the entity(ies) to turn *cyan*, add the appropriate relation(s) in the "Add Relations" menu, and

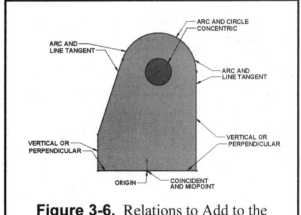

Figure 3-6. Relations to Add to the Sketch.

then click the green check mark (√). Repeat these steps to add the following as indicated in **Figure 3-6**.

- Origin and Bottom Horizontal Line are **Coincident** and **Midpoint**.

- Right Side line is **Vertical** or **Perpendicular**.

- Right Side line and Top Arc are **Tangent**.

- Top Arc and Circle are **Concentric**.

- Top Arc and Left Angled Line are **Tangent**.

This completes all the relations for the sketch. However, your current sketch may not look like the final version. It may be too short or too tall or too wide. Also your lines are *blue* meaning the geometry is under-defined. Try something now. With the **Select** cursor, pick the center of the concentric circle and arc. Hold the **LMB** down and move it up, to the right, or to the left. See the sketch change shape, but the relations (like tangent, vertical, midpoint, and perpendicular) remain true and fixed. This is called constraint-based modeling. Now return the center point back to its original position.

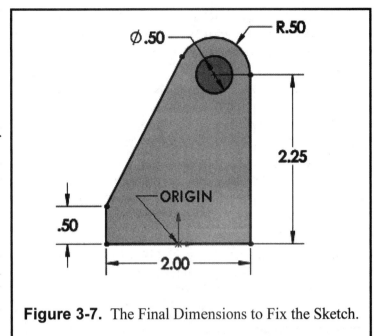

Figure 3-7. The Final Dimensions to Fix the Sketch.

You will now complete the geometry of your sketch by adding dimensions. Study the dimensions to add in **Figure 3-7**. With the **Dimension** tool, apply the dimensions (all in inches). Start with the **2.25** height of the center of the concentric circle and arc. Then apply equal **0.500** diameter and radius dimensions for the circular features. Finally, add the **2.000** length of the bottom line and the **0.50** height of the small vertical line on the left side. Now the sketch, with relations and dimensions added, is fixed and all lines turn *black*.

You can now try another thing. With the **Select** cursor, double click the **2.000** dimension of the bottom line and change it temporarily to **4.000.** Notice how wide the base becomes, but it is centered at the origin because of the "Midpoint" relation it possesses. Then change that same dimension to **1.000.** See how narrow it becomes, but the "Midpoint" and "Tangent" relations still hold true and stand out in the design. Now return the dimension back to the original value of **2.000.** You are now ready to extrude the sketch to get some 3D geometry for the Clevis Mounting Bracket.

Click the **Extrude Boss/Base** icon on the "Features" toolbar and the "Base-Extrude" menu appears. Give it a **Blind** extrusion distance of **0.375** inches toward the "**Original Right Plane**" (see **Figure 3-8**). You can use the flip direction button if needed to get it pointed towards that *Right* plane **(away from your point of view)**. Use an **Isometric** view to aid in seeing this direction and then click the green mark (√). You should now have a solid piece of the Bracket as shown in **Figure 3-8**.

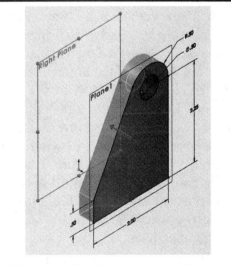

Figure 3-8. The Extruded Sketch.

Now return to a **Right** view and click the front surface (it turns *blue*) to sketch on it. Click the **Sketch Tab** and **Select** the **Sketch Icon**. Pick the **bottom horizontal edge**, then hold down the **Ctrl** key and pick the **left short vertical edge** of the model (they turn *cyan*). Next, click the **Convert** sketch icon to convert *those edges* into lines for the current sketch. They still retain the previous data so they are *black* lines. Now draw a **Line**, from the top of the converted vertical line, horizontally over to the right edge. Then draw a final **Line** vertically down to the bottom right corner. This essentially makes a rectangular sketch that can be extruded to form the bottom of the Clevis Mounting Bracket. Click the **Extrude Boss/Base** icon on the "Features" toolbar. Give it a **Blind** extrusion distance of **2.750** inches <u>back</u> into the direction of the "**Original Right Plane**" (use flip direction if needed). Use an **Isometric** view to aid in seeing this direction. Click the green check (√) to complete this bottom base.

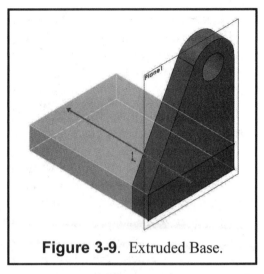

Figure 3-9. Extruded Base.

Now you need to create a symmetric feature like **Figure 3-8** on the backside of the model. Hold down the **Ctrl** key, and on the Feature Manager tree, click both the **Right** plane label and the **Base-Extrude 1** label. Now click the **Mirror Feature** icon on the "Features" toolbar (see icon to right). The "Base-Extrude" upright feature is now mirrored about the front plane, as shown in **Figure 3-10**.

[icon: Mirror / Mirror Feature]

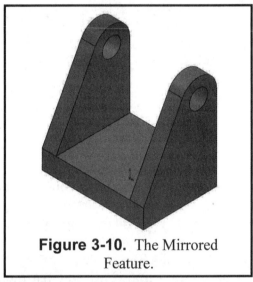

Figure 3-10. The Mirrored Feature.

You will now complete the Bracket by adding some features to the bottom base. Refer to **Figure 3-11** for this data. Select a **Top** view orientation. Click on the top surface of the bottom base to turn it *blue* and then click the **Sketch** icon. Use the **Circle**, **Arc** and **Line** tools to draw the hole and slot outline. Use the **Dimension** tool to add the given dimensions. The slot at the bottom can extend over the edge a little to make sure it cuts off the solid middle part of that edge. **Add a relation so the center point of the circle and the arc are aligned vertically with the origin**. If it is correct, the lines should turn *black* once all dimensions are applied, except for the portion extending beyond the edge.

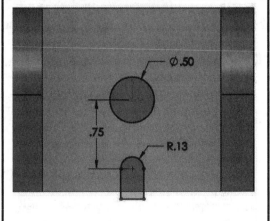

Figure 3-11. Geometry for the Bottom Holes.

Now click the **Extruded Cut** icon. In the "Extruded-Cut" menu, give it the **Through All** condition and make sure the cut direction is downward. Use an **Isometric** view to aid in visualizing this operation. Then click the green check mark (√). One last operation to complete the solid model is to add a **0.125 Fillet** to the edges where the vertical surfaces intersect the horizontal surface. Your Clevis Mounting Bracket is now complete. View your final model in an **Isometric** orientation as shown in **Figure 3-12**. The **CLEVIS MOUNTING BRACKET** is made of Bronze material properties that can be reflected by the color of the shaded image. In the Feature Manager Tree, right click on **Edit Material**, expand the **Copper Alloy** materials category and assign **Leaded Commercial Bronze** to the **CLEVIS MOUNTING BRACKET**.

Pull down **File**, select **Save As**, select your proper folder, type in the part name **CLEVIS MOUNTING BRACKET.sldprt**, and then click **Save**. Insert the rendered Clevis Mounting Bracket image onto your **Title Block** drawing sheet, as shown in **Figure 3-12**. Now **SAVE** your drawing sheet to your directory as **CLEVIS MOUNTING BRACKET.slddrw**. Take note that the title of the part and the drawing are the same but the extension has changed. The solid model and the drawing are linked, meaning that any changes made to the solid model will automatically be updated on the drawing (see **Figure 3-12**).

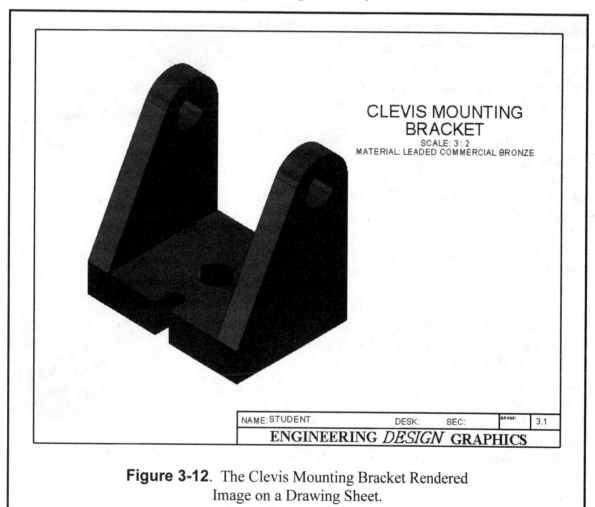

CLEVIS MOUNTING
BRACKET
SCALE: 3 : 2
MATERIAL: LEADED COMMERCIAL BRONZE

| NAME: STUDENT | | DESK: | SEC: | DR SET: | 3.1 |

ENGINEERING *DESIGN* GRAPHICS

Figure 3-12. The Clevis Mounting Bracket Rendered
Image on a Drawing Sheet.

Exercise 3.2: MANIFOLD

In this Exercise 3.2, you will design a Manifold. The Manifold is designed to allow air or fluid to flow in many directions through its ports. There will be many similar flow ports, so you will use some 3D editing commands in SOLIDWORKS to replicate them.

Go to your folder and **Open** the file **ANSI-METRIC.prtdot** since the Manifold is designed in metric units. In order to avoid corrupting the **ANSI-METRIC.prtdot** template, go to **File - Save as:** under **File Name**, type **MANIFOLD**, and under **Save as type**, select **.sldprt** and select **SAVE.** Click the **Right** plane icon in the Feature Manager, select a **Right** view orientation and select the **Sketch Tab**, then select **Sketch** to start the sketch. Draw two **Circles** that are concentric and centered at the **ORIGIN**, with one circle a little larger. **Dimension** the larger circle to be **60** mm in **diameter** and the smaller circle to have a **45** mm **diameter**. These two dimensions should fix your geometry.

Click the **Features Tab** and select **Extrude Boss/Base**. In the "Base-Extrude" menu, pick **Direction 1** to be **Mid Plane** at **300 mm**, then click the green check mark (√). You now have a long throat centered at the origin, as shown in **Figure 3-13**.

Now add a collar on the visible end. Click on the right side surface of the throat hole (defined by the two concentric circles). The surface should turn *blue*. Click on the **Sketch Tab** and select the **Sketch Icon. Pick** the smaller inner circular edge. Click the **Convert** icon to convert it to the new sketch. Now draw one **Circle**, centered at the origin, and **Dimension** it with a diameter of **75** mm.

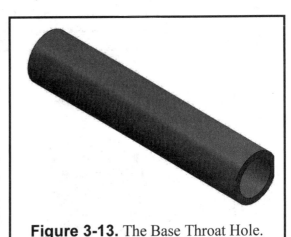

Figure 3-13. The Base Throat Hole.

Figure 3-14. Adding Collars to the Ends.

Click the **Extrude Boss/Base** icon. In the "Base-Extrude" menu, pick **Direction 1 (back over the throat)** to be **Blind** at **50** mm then click the green check mark (√). You now have a small collar on one end of the throat. You can easily copy it to the other side. Hold down the **Ctrl** key. In the Feature Manager, click both the **Right** plane icon and the last **Boss-Extrude 2**. They both should turn *cyan* in the computer modeling space. Go to the **Features** and select the **Mirror** feature icon on the features toolbar. Click the green check mark (√) in the "Mirror Feature" menu to execute the operation. You now have a collar on each end of the throat, as shown in **Figure 3-15**.

You will design the first port hole. Click the **Top** plane in the Feature Manager. Now Pull down **Insert**, select **Reference Geometry,** and then select **Plane**. The "Plane" menu appears. Indicate the distance to be **42.5** mm above the top plane, then click the green check mark (√) to add the plane to the working space. **Plane 1** is added to the Feature Manager, so click on it (it turns *cyan*) and start another **Sketch**. Click a **Top** view orientation to better see this new sketching plane.

Now draw a small **Circle** on the plane just inside the collar on the left side of the throat. Click the **Add Relations** icon, and click both the <u>center</u> of the circle and the origin. Under the "Add Relations" menu, select the **Horizontal** relation to these two entities. Click the green check mark (√) to add this relation to make them align horizontally. Next, **Dimension** the circle to have a diameter of **35** mm and to be **75** mm to the left of the origin, as shown in **Figure 3-15**.

Now click the **Extrude Boss/Base** icon in the features toolbar on the left side of the screen. In the "Base-Extrude" menu, pick Direction 1 to be **Up to Surface.** Pick the outer surface of the main body of the model (it turns *pink*), then click the green check mark (√). You now have a boss extruded down from Plane 1 to the outer surface.

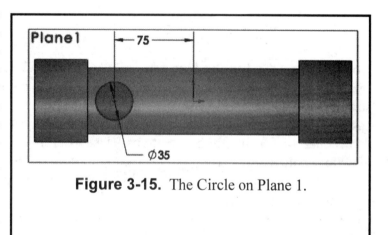

Figure 3-15. The Circle on Plane 1.

Now repeat a **Sketch** on **Plane 1**, and draw a smaller **Circle** that is concentric with the boss (**Add Relations, Concentric**) and has a diameter of **17.5** mm.

Return to an **Isometric** view orientation. Click the **Extruded Cut** icon on the Features Tab. In the "Cut-Extrude" menu, pick Direction 1 to be **Up to Surface.** Pick the *inner throat surface* on the model (it turns *pink*) by clicking inside the throat hole on the right side. Click the green check mark (√). You now have a hole extruded down from Plane 1 into the inside of the throat as shown in **Figure 3-16**.

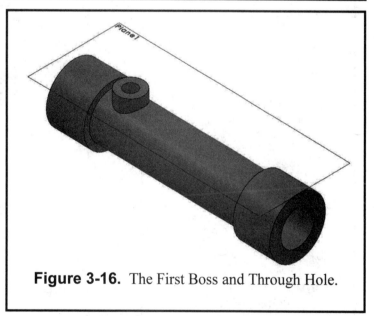

Figure 3-16. The First Boss and Through Hole.

To better see the through hole, use the **Rotate View** icon to rotate your model around the computer screen. Try to view the through hole by peering down the throat of the part. Later in this exercise you will cut a temporary section view to see things on the inside better.

Now that you have one boss and hole feature, it is easy to create a 3D array of the other three that are needed. First you need an axis for the direction of the 3D array. Pull down **Insert**, select **Reference Geometry,** and then select **Axis**. The "Reference Axis" menu appears on the screen as shown in **Figure 3-17**. Select the **Cylindrical/Conical Face** definition for the axis and then click the outer surface (**Face<1>**) of the main cylinder. It gets added to the "Selections" list and then click the green check mark (√). You now have added an "**Axis 1**" to the Feature Manager tree that goes down the center of the throat. Press down the **Ctrl** key, and one-by-one select in the Feature Manager tree:

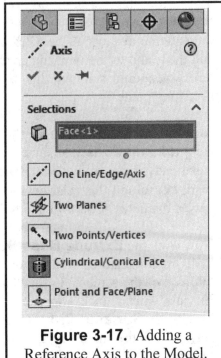

Figure 3-17. Adding a Reference Axis to the Model.

Axis 1
Boss-Extrude 3
Cut-Extrude 1

Now click on the **Linear Pattern** icon of the **FEATURE MENU**. The "Linear Pattern" menu now appears on the screen as shown in **3-18**. There is only one Direction for this case. The parameters for "Direction 1" should be along **Axis 1**, distance of **50** mm, and **4** patterns. Use your Isometric view to see the patterns replicate over the Manifold to the right side. Then click the green check mark (√). You should now have a pattern of four port holes as shown in **Figure 3-19**. Try viewing down the middle of the throat with the **Rotate View** icon.

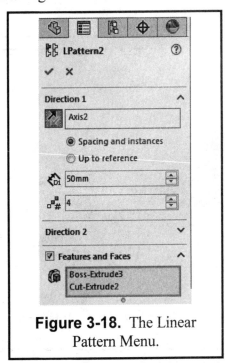

Figure 3-18. The Linear Pattern Menu.

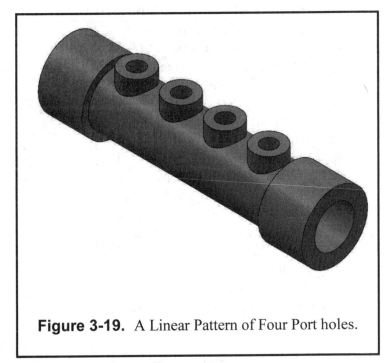

Figure 3-19. A Linear Pattern of Four Port holes.

There are two remaining port holes that need to be added to the bottom of the Manifold. They are identical in geometry to the top, so you can just mirror the port hole in 3D space. Hold down the **Ctrl** key and, in the Feature Manager tree, one-by-one select **Top Plane**, **Boss-Extrude 3**, and **Cut-Extrude 1.** Then click the **Mirror Feature** icon and click the green check mark (√) in the "Mirror Feature" menu to execute the operation. You now have one port hole on the bottom of the Manifold, which is named "Mirror 2." Repeat this 3D **Mirror Feature** operation, this

time using the **Right Plane** and **Mirror 2.** After you complete this operation, you will have a "Mirror 3" feature for the Manifold solid model, as shown in **Figure 3-20**.

Now add some solid feature fillets to the edge where the portholes touch the mail body. Click the **Fillet** icon on the **FEATURES MENU**. Enter the fillet radius of **6 mm** in the "Fillet" menu. One-by-

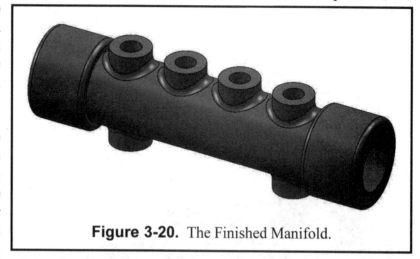

Figure 3-20. The Finished Manifold.

one click on the edges (loops) of the six port holes where they touch the outer surface of the throat (they turn cyan). You can do this easily with a **Front** view orientation. Then click the green (√) to execute the fillet command.

With the **4mm** radius **Fillet** the sharp edges of the two collars. See the finished **Figure 3-20** in **Trimetric** view.

Earlier you rotated the model around to see through the long throat hole to see if the port holes went through to the middle. Now you will create a temporary section view to see internal features. Pull down **View**, select **Display**, and then pick **Section View**. A "Section View" menu appears on the screen. The section view parameters should be set to **0** mm distance and the **Front** plane is the default plane, as shown in **Figure 3-21**. Just press cancel (**X**) on the menu, after you have inspected the inside features of the manifold.

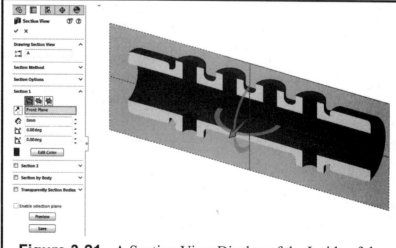

Figure 3-21. A Section View Display of the Inside of the Manifold.

The **MANIFOLD** is made of TITANIUM material properties that can be reflected by the color of the shaded image. In the Feature Manager Tree, right click on **Edit Material**, expand the **OTHER METALS** materials category and assign **TITANIUM** to the **MANIFOLD.**

Pull down **File,** select **Save As,** type in the part name **MANIFOLD.sldprt**, select your proper folder, and then click **Save**. Insert the image into the **Title Block** drawing sheet like in previous labs (refer to page 1.7).

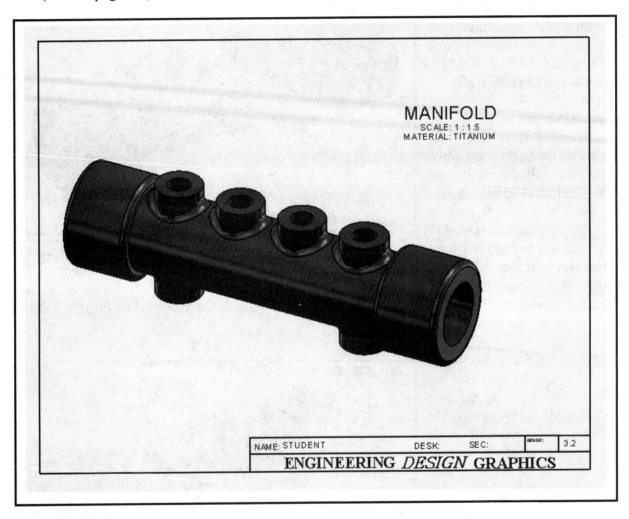

MANIFOLD
SCALE: 1 : 1.5
MATERIAL: TITANIUM

NAME: STUDENT		DESK:	SEC:	GRADE:	3.2
ENGINEERING *DESIGN* GRAPHICS					

Exercise 3.3: HAND WHEEL

In this Exercise 3.3, you will design a Hand Wheel. The Hand Wheel has an elliptical cross-section that can be revolved 360 degrees. You will learn how to sketch an ellipse. The five spokes for the hand wheel can be created using a 3-D circular pattern array. You will employ some new SOLIDWORKS commands in this lab exercise.

Go to your folder and **Open** the file **ANSI-INCHES.prtdot**. Immediately go to **File - Save as:** under **File Name**, type **HAND WHEEL**, and under **Save as type**, select **.sldprt** and select **SAVE**. Click on the **Front** plane in the Feature Manager tree. Set the view orientation to **Front**, then start a new **Sketch**. If you do not have an ellipse icon, you need to go to **Tools**, select **Sketch Entity**, and then select **Ellipse**. Draw an ellipse off to the right side of the origin as shown in **Figure 3-22**. Click the **LMB** cursor to place the center of the ellipse off to the right side of the origin. Drag **vertically** above the center and click the **LMB** to set the major axis of the ellipse. Then drag horizontally from the center and click the **LMB** again to set the minor axis of the ellipse.

Right now the ellipse is somewhat randomly placed on the frontal sketch plane. The center of the ellipse should be aligned with the origin. Click on the **Display/Delete Relations down arrow** and select the **Add Relations** icon. Pick the **center of the ellipse** and then pick the **Origin**. Add the **Horizontal** relation to align them and then click the green check mark (√). Now **Dimension** the ellipse using **Figure 3-22** as a guide. The major diameter is **1.000** inches, the minor diameter is **0.750** inches, and the center's distance from the origin is **4.500** inches. Also sketch a **vertical Centerline** through the origin as shown in **Figure 3-22**. The ellipse is now ready to be revolved.

Click the **Revolve Boss/Base** icon on the features toolbar. In the "Base-Revolve" menu, set the menu parameters to **One Direction** and to **360** degrees, then click the green check mark (√). You now have the first feature for the Hand Wheel as shown in **Figure 3-23**.

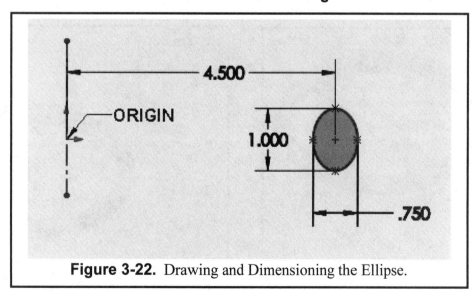

Figure 3-22. Drawing and Dimensioning the Ellipse.

You will now construct a spoke from the center of the Hand Wheel. It also has an elliptical shape as shown in **Figure 3-24**. Click the **Front Plane** in the Feature Manager and select a **Front** view orientation. **Sketch** an **Ellipse** centered at the origin. Next, **Dimension** the ellipse according to **Figure 3-24**.

Switch to an **Isometric** view orientation to better see the next operation. Click the **Extrude Boss/Base** icon and select the **Up to Surface** end condition for Direction 1. Now **Select** the round tube surface of the Hand Wheel (**Figure 3-23**) for the surface for the end condition for the spoke. It should turn *pink* and the extruded spoke should preview on the screen. If it looks correct, click the green check mark (√) and the operation is executed. You should now have one spoke as shown in **Figure 3-25**.

You can now add the remaining spokes using a circular pattern feature. But first you need to add an axis for the pattern function. If the origin is not visible, pull down the **View – Hide/Show** menu and select **Origins**. Pull down **Insert**, **Reference Geometry**, and select **Axis**. The "Reference Axis" menu appears on the screen as shown in **Figure 3-26**. Now carefully **Select** the **Origin** as the "Point." Next, **Select Two Planes** as the axis option. Using the Feature Manager to the right of the Axis Window, select the **Front Plane** and the **Right Plane**. Then click **OK** to close the menu. You should now see an Axis 1 on your screen.

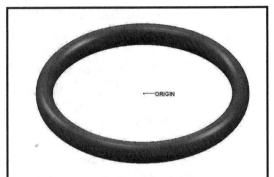

Figure 3-23. The First Feature of the Hand Wheel.

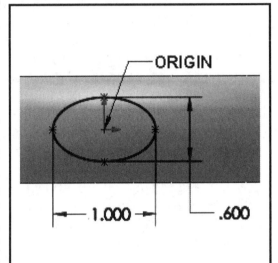

Figure 3-24. Dimensions for the Elliptical Spoke.

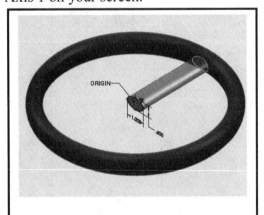

Figure 3-25. The Extruded Spoke.

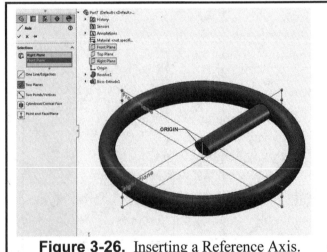

Figure 3-26. Inserting a Reference Axis.

Hold down the **Ctrl** key and **Select** both the **Axis 1** (you can click it in the Feature Manager area) and the **Spoke**. They should highlight in *blue*. Click on the **Circular Pattern** icon on the Features Toolbar and the "Circular Pattern" menu appears. Since they were pre-selected, **Axis 1** should appear in the "Parameters" box and **Boss-Extrude 1** should appear in the "Features to Pattern" box. Now set the "Angle" to **360** degrees and the number of patterns to **5**. Now click the green check mark (√) and the circular pattern with the spokes is created, as shown in **Figure 3-27**.

Figure 3-27. Circular Pattern of Spokes.

Now you will add the central hub of the Hand Wheel. Select the **Top** plane in the Feature Menu and select a **Top** view orientation. Click **Sketch Tab** and **Select** the **Sketch Icon** and draw a **Circle** centered at the origin with a diameter **Dimension** of **2.000** inches. Switch to **Isometric** view orientation. Click **Extrude Boss/Base** features icon. On the "Extrude-Boss" menu, set the following parameters:

Direction 1: **Mid Plane, 1.50** inches

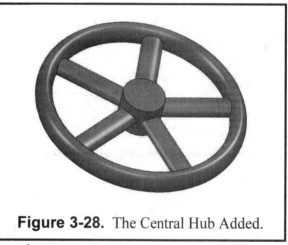

Figure 3-28. The Central Hub Added.

Place a check mark (√) in the box labeled "**Merge Result**." Now click the green check mark (√) to close the menu. The center boss hub is extruded in both directions as shown in **Figure 3-28**.

You will now add the hole and keyway to go through the boss. Click on the top surface of the central boss to highlight it in *blue*. Also select a **Top** view orientation. Now click the **Sketch** icon and draw a sketch similar to the one shown in **Figure 3-29**. You will need to draw a **Circle** and then either a few **Lines** or maybe a **Rectangle**. Use the **Trim** sketch editing capability to cut away the profile for the keyway. Then **Dimension** the sketch as shown in **Figure 3-29**:

Circle Diameter = **1.25** inches
Keyway width = **0.220** inches
Keyway height = **0.750** inches
Keyway distance = **0.110** inches

If all is correct, your sketch should turn *black* meaning the geometry is fixed.

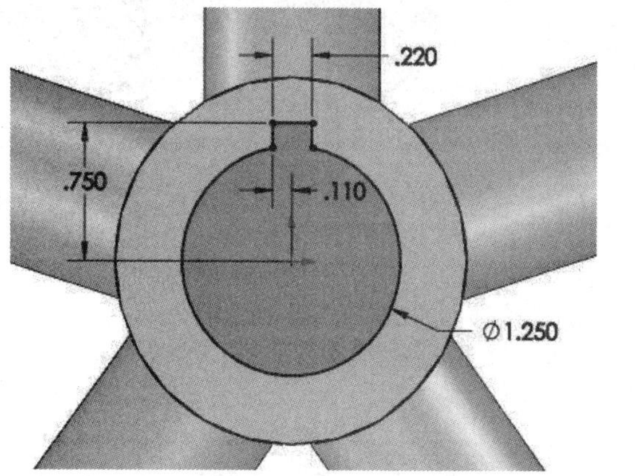

Figure 3-29. The Sketch and Dimensions for the Through Hole and Keyway.

Now click the **Extrude Cut** icon and select the **Through All** end condition. Make sure the direction for the "through all" condition is down, then click the green check mark (√). The cut is created through the hub as shown in **Figure 3-30**. The Hand Wheel is almost complete; you just need to add the fillets to the places where the spokes intersect the hub and wheel hub.

Now click the **Fillet** features icon on the Features toolbar (it looks like a block with a round corner). Key in a fillet radius of **0.125** inches. One-by-one, pick the five surfaces of the five spokes to fillet the edges. They turn *blue* when picked and they get added to the "Items to Fillet" list in the menu. Then click the check mark (√) to execute the ten fillets. The finished Hand Wheel model is shown in **Figure 3-30** in an **Isometric** view.

Figure 3-30. The Finished Hand Wheel Model Shown in Isometric.

Pull down **File,** select **Save As,** type in the part name **HAND WHEEL.sldprt,** select your proper folder, and then click **Save.** Insert the image into the **Title Block** drawing sheet like in previous labs (see **Figure 3-31**). **Print** a hard copy to submit to your lab instructor. Save your drawing as **HAND WHEEL.slddrw.**

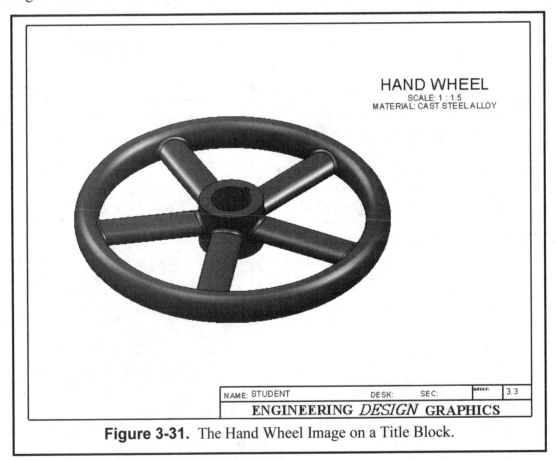

Figure 3-31. The Hand Wheel Image on a Title Block.

Exercise 3.4: TOE CLAMP

In the previous lab exercise, you created sketches on planes that were orthogonal to the principal planes (Front, Top, or Right). In this simple Exercise 3.4, you will create features for the Toe Clamp by sketching on both orthogonal and inclined planes. The inclined plane is at a 45 degree angle to the top surface, so you will also learn how to dimension angles.

Go to your folder and **Open** the file **ANSI-INCHES.prtdot**. Immediately go to **File - Save as:** under **File Name**, type **TOE CLAMP**, and under **Save as type**, select **.sldprt** and select **SAVE.** Click on the **Front** plane in the Feature Manager tree. Set the view orientation to **Front**, then start a new **Sketch**. You need to draw the front outline of the Toe Clamp as shown in **Figure 3-32**. Use the **Line** and **Dimension** tools to create and fix the geometry. **Add Relations** to the bottom **Horizontal** line and make it both **Coincident** and **Midpoint** with the origin. When dimensioning an angle, pick the two lines that form the angle and then drag the angle dimension into place.

Now select the **Extrude Boss/Base** icon to create the base part. Enter the following parameters in the "Base-Extrude" menu:

Direction 1:

Mid Plane, 2.00 inches

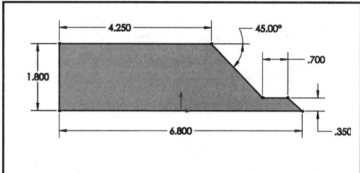

Then click the green check mark (√) to execute the bi-directional extrude, as shown in **Figure 3-33**.

Figure 3-32. The Initial Shape and Dimensions for the Toe Clamp.

You will now cut the counterbore and hole on the inclined surface. Click on the inclined surface to highlight it *blue*. Now you want to see this surface head on (ortho-directional). So click the **Normal to** selection in the View Orientation box (or you can also click the **Normal to** icon on the top display toolbars as shown below).

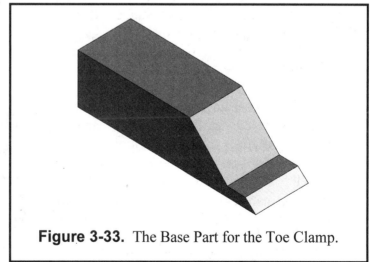

Figure 3-33. The Base Part for the Toe Clamp.

Sketch a **Circle** on the inclined plane as shown in **Figure 3-34**. **Dimension** the diameter to be **0.50** inches. Center the circle **1.00** inches vertically. Add a **vertical** relationship between the center of the circle and the origin. Click the **Extruded Cut** icon on the features toolbar and give it the **Through all** end condition. Make sure the hole is cut at an angled, downward direction and then click the green check mark (√). You can see it with the **Rotate View** icon.

Click on the inclined surface and return to a **Normal to** view of it, and then **Sketch** another larger **Circle** on it. **Dimension** it with a diameter of **0.80** inches. Next, click **Add Relations** and make this new circle **Concentric** with the edge of the through hole. Click the **Extruded Cut** icon and give it a **Blind** end condition of **0.25** inches into the part. You now have a counterbore hole as seen in **Figure 3-35**.

You can now make the V-cut to add stress relief for the Toe Clamp. Click on the plateau horizontal surface just below the inclined surface. Change your viewpoint to the top view. Click the **Sketch tab** and **Select** the **Sketch** icon; with the **Line** tool draw a small triangle shape approximately over the beveled edge of the Toe Clamp, as suggested in **Figure 3-36**. Apply **Dimensions** as shown in the Figure. If you apply them correctly, the sketch will turn *black*.

Now you can execute the V-Cut. Click the **Extruded Cut** icon on the features toolbar and give it the **Through all** end condition. Make sure the cut goes in the downward direction through the beveled end, and then click the green check mark (√). Use the **Rotate View** icon to see it.

From the **Feature Manager Tree - Select** the **Right Plane then the Right View** and then select the **Sketch Icon**.

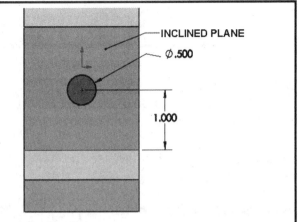

Figure 3-34. Creating the Circle on the Inclined Plane.

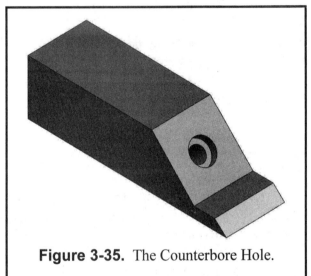

Figure 3-35. The Counterbore Hole.

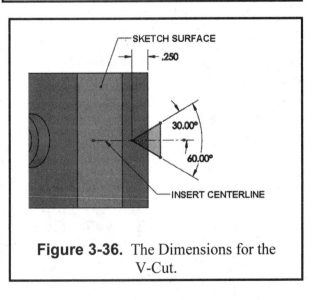

Figure 3-36. The Dimensions for the V-Cut.

3-18

Now Select the Centerline command and **Draw** a **Vertical Centerline** through the **Origin**. Next, go to **Tools – Sketch Tools - Dynamic Mirror**. Your next step is to **Select** the **Rectangle** and Sketch a Rectangle along the **Left Edge** of your model as shown in **Figure 3-37**. This rectangle will be duplicated on the Right Side because of the Dynamic Mirror Command. Activate the Features Tab and execute an **Extruded Cut – Through All** in both Directions. The resulting model will look like **Figure 3-39**.

The final feature to create is the large counter slot on the very top surface of the Toe Clamp. Click on the top surface (it turns *blue*) and **Sketch** the outer slot shape as shown in **Figure 3-40**. Use the **Straight Slot** sketch tool to begin the sketch. **Add Relation** as follows:

The centers of the arcs are to be **Horizontal** with the **ORIGIN.**

Next, **Dimension** the slot by applying the values given in **Figure 3-40**. The sketch should be fixed and turn *black*. Click the **Extruded Cut** icon and give it a **Blind** end condition of **0.25** inches into the part. The 0.25 deep counter slot is created.

Select the same top surface to start the final sketch. Activate the **Sketch** tab and select **Sketch**. Identify the large slot edge you just drew, and **Convert** it. Next, **Offset** it inward by **0.20** inches to create a smaller slot that has the identical shape. **Delete** the bigger slot on the sketch. Click the **Extruded Cut** icon and give it the **Through all** end condition. Make sure the slot is cut in a downward direction, and then click the green check mark (√). (*Note:* This process for creating the counter slot was the reverse of the counterbore process, but they both worked correctly.) The Toe Clamp model is now finished as shown in **Figure 3-41** in an **Isometric** view.

The **TOE CLAMP** is made of Bronze material properties that can be reflected by the color of the shaded image. In the Feature Manager Tree, right click on **Edit Material**, expand the **Copper Alloy** materials category and assign **Leaded Commercial Bronze** to the

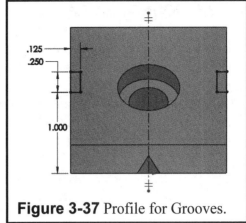

Figure 3-37 Profile for Grooves.

Figure 3-38 Grooves of the Toe Block.

Figure 3-39 Grooved Toe Block.

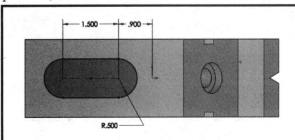

Figure 3-40. The Counter Slot Feature.

CLEVIS MOUNTING BRACKET.

Pull down **File,** select **Save As,** type in the part name **TOE CLAMP.sldprt**, select your proper folder, and then click **Save**. Insert the image into the **Title Block** drawing sheet like in previous labs. Save your drawing as **TOE CLAMP.slddrw**. **Print** a hard copy to submit to your lab instructor.

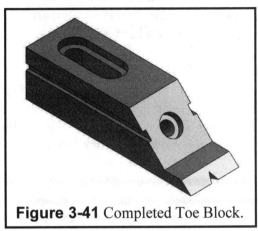

Figure 3-41 Completed Toe Block.

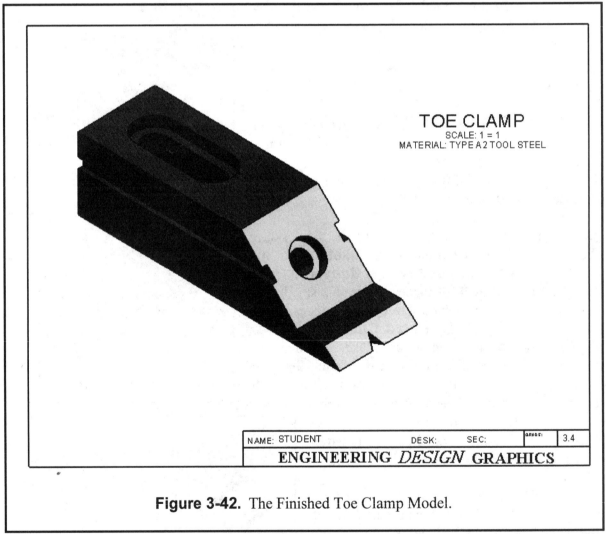

TOE CLAMP
SCALE: 1 = 1
MATERIAL: TYPE A2 TOOL STEEL

NAME: STUDENT   DESK: SEC: GRADE: 3.4

ENGINEERING *DESIGN* GRAPHICS

Figure 3-42. The Finished Toe Clamp Model.

SUPPLEMENTARY EXERCISE 3-5: CONVEYOR RAMP GUIDE

Build a full size model of the figure below. Insert it on a Title Block and title it **"CONVEYOR RAMP GUIDE."**

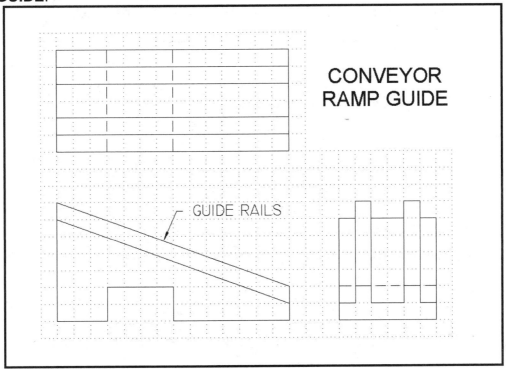

ASSUME THE GRID DIVISIONS TO BE 0.25 INCHES.

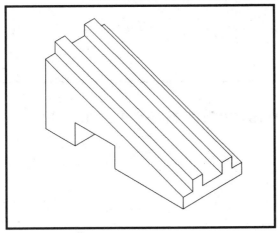

PICTORIAL VIEW

SUPPLEMENTARY EXERCISE 3-6: DOUBLE SHAFT HANGER

Using the commands learned during the last two Units, build a full size model of the figure below. Insert it onto a Title Block and title it **DOUBLE SHAFT HANGER**.

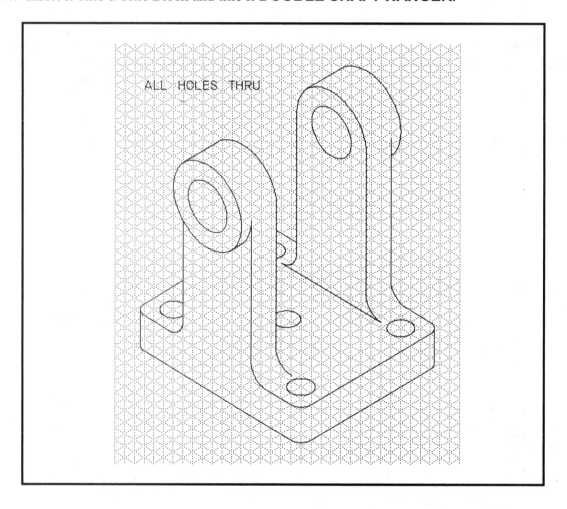

ALL HOLES THRU

ASSUME THE GRID DIVISIONS TO BE 0.25 INCHES.

Computer Graphics Lab 4:
3-D Solid Modeling of Parts II

INTRODUCTION TO LAB 4

In this lab you will be using many of the design features that make building solid models more efficient. Commands you will be using:

Draft - Creates a feature that tapers selected model faces by a specified angle, using either a neutral plane or a parting line. **NOTE:** You can also apply a draft angle as a part of an extruded base, boss, or cut.

Offset Plane - You can create planes in part or assembly documents. You can use planes to sketch, to create a section view of a model, for a neutral plane in a draft feature, and so on.

Offset - You can create sketch curves offset from one or more selected sketch entities, edges, loops, faces, curves, set of edges, or set of curves by a specified distance. The selected sketch entity can be construction geometry. The offset entities can be bi-directional.

Convert Entities - You can create one or more curves in a sketch by projecting an edge, loop, face, curve, or external sketch contour, set of edges, or set of sketch curves onto the sketch plane. You can convert sketched entities into construction geometry to use in creating model geometry.

Fillet surfaces - Fillet/Round creates a rounded internal or external face on the part. You can fillet all edges of a face, selected sets of faces, selected edges, or edge loops.

Shell - The shell tool hollows out the part, leaves open the faces you select, and creates thin-walled features on the remaining faces.

Loft - Creates a feature by making transitions between profiles. A loft can be a base, boss, cut, or surface.

Dome - Creates a loft type of feature that begins with the shape of the selected surface and lofts to a zero feature at a specified height.

Sweep - Creates a base, boss, cut, or surface by moving a profile (section) along a path, according to these rules:
- The profile must be closed for a base or boss sweep feature; the profile may be open or closed for a surface sweep feature.
- The path may be open or closed.
- The path may be a set of sketched curves contained in one sketch, a curve, or a set of model edges.
- The start point of the path must lie on the plane of the profile.
- The section, the path, or the resulting solid cannot be self-intersecting.

Exercise 4.1: DRAWER TRAY

Go to your folder and open **ANSI-INCHES.prtdot**. A good practice is to immediately name your part and save the file in your folder. Pull down **File** again, select **Save As**, and then name it **DRAWER TRAY.sldprt**.

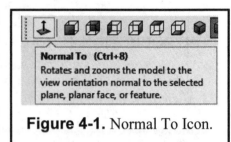

Figure 4-1. Normal To Icon.

Select the **Top Plane** in which to work, then click on the **Normal To** icon (**Figure 4-1**) which looks like a square plate with an arrow pointing upwards out of the center of the plate.
Select the **Sketch Tab** and activate the **Sketch Icon,** then draw a **Rectangle** at the origin and toward the upper right.

Using the Dimension command, constrain the rectangle to **4.5** inches high and **12** inches long. Now **Fillet** the four corners with a radius of **.375"**. The resulting sketch should look like the object in **Figure 4-2**.

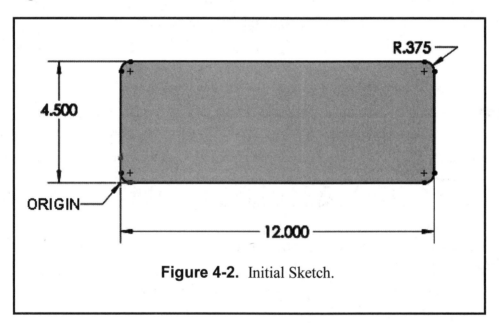

Figure 4-2. Initial Sketch.

Select the **Extrude** symbol (**Figure 4-3**) under Features.

Do a **Blind Extrude downward** for **3** inches, with a **Draft Angle** of **5 Degrees** (See **Figure 4-4**). If the upward-pointing arrow is highlighted, select it and pull it downward or click on the "Arrow" icon next to the "**Blind**" box. Click on the (√) button, which will produce an image similar to **Figure 4-5**.

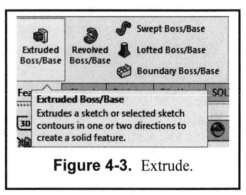

Figure 4-3. Extrude.

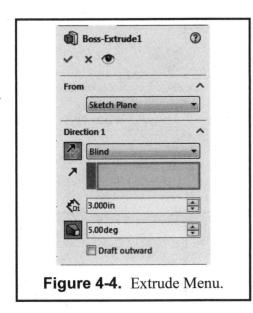

Figure 4-4. Extrude Menu.

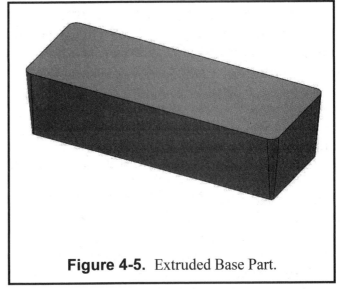

Figure 4-5. Extruded Base Part.

Click on the words "**Base Extrude**" in the Feature Manager twice (this is not a double click) and rename this feature as "**Tray Body**."

Now rotate 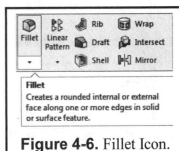 the tray so you can see the bottom surface. *If you press down the center wheel of your mouse, you have the rotate command.* Select the bottom surface and under the **Feature Tab** pick the solid **Fillet** feature (**Figure 4-11**) from the **Features** tool bar. Enter **0.3750** inches. The entire outline or loop of this surface will be filleted.

Figure 4-6. Fillet Icon.

Click on the **Top** surface of the Tray Body then pick the **Shell** icon (**Figure 4-7**) from the **Features Tab** in the upper left tool bar. Give the shell thickness of **0.10** in the parameters box and okay the selection. The model should now resemble the model in **Figure 4-8**.

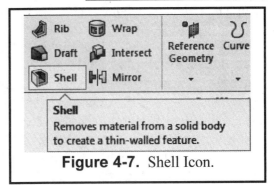

Figure 4-7. Shell Icon.

Drawer Tray Rim. In order to strengthen the upper edge of the drawer tray you will now create a rim around the top edge of the tray.

You might have to **Zoom** in to select the top thin surface of the resulting tray (use the center wheel of your mouse). After you have selected the thin surface, **select** the **Sketch Tab** then activate the **Sketch Icon**. Now select the **Convert Entities** icon (**Figure 4-9**). This converts the outer edges of that surface into new sketch entities that can be used for a new feature. Next,

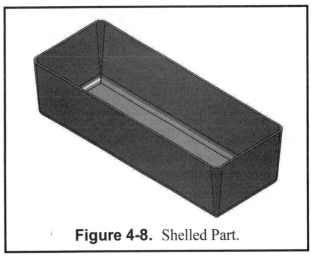

Figure 4-8. Shelled Part.

select the **Offset Entities** icon (see **Figure 4-9**) and pick one of the black edges of the converted entities; enter the distance of **0.10**. If the yellow preview line is on the outside of the pattern selected, then select

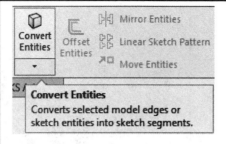

Figure 4-9. Convert Entities. Icon.

Figure 4-10. Offset Icon.

the **Reverse** option (**Figure 4-10**) so the new offset pattern lies directly over the entire top edge surface of the Drawer Tray. Extrude this pattern downward by **0.625** inches. Make sure that the **Merge Result** box is checked. This creates a vertical rim around the upper edge of the tray.

The next step of the construction is to place three equally spaced dividers on the inside of the tray. Select the **Top Plane** in the feature manager. Then select **Insert, Reference Geometry, Plane**. Enter the distance of **0.25** and make sure the **Flip** box is checked (**Figure 4-11**). Click on the green (**OK**) button. This establishes a new sketch plane below the Top Plane that can now be used to draw the top view of the ribs. It might be advisable to rename this plane. Rename it **Dividers Plane**.

Figure 4-11.
Reference Plane Menu.

For the next step it might be advisable to turn off the hidden lines. To do this go to **Display Style** and **Select** the **Hidden Lines Removed (Figure 4-12)**. Select the **Dividers Plane** in the Feature Manager and then the **Normal To** icon to get a top view of the tray. Next, select the **Sketch** icon and draw a **Rectangle** feature **across the inside edges** of the tray. Constrain the rectangle to be **0.10** inches across and **3** inches from the outside of the Tray. See **Figure 4-13**. Select the **Linear Sketch Pattern** icon, activate the "**Entities to Pattern**" box and select the four lines of the rectangle. Supply the following criteria: **Number (3); Spacing (3); and Angle (180)**. If the preview of the array does not show the repeated pattern on the Tray, click on the icon next to the **Direction 1** input box (**Figure 4-14**) to change the direction of the array. If the preview is correct, select the **OK** button.

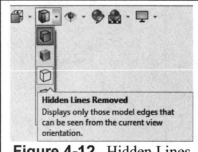

Figure 4-12. Hidden Lines removed.

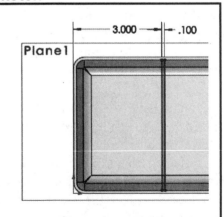

Figure 4-13. Draw a Rectangle.

You are now ready to extrude the patterns downward to the Tray Shell. Select the **Features** tab, and **Select Extruded Boss/Base.** The extrusion arrow will be pointing upward. Click on the highlighted arrow and pull it downward. Then go to the **"Boss Extrude" menu** and select the **Up To Next** option (**Figure 4-15**) and click **OK**.

Right click on **Material (not specified)** in the Feature Manager. Select **Edit Material**. Expand the **Plastics Tab** and select **PC High Viscosity** and then select **Apply** and **Close**. Your completed model should look like **Figure 4-16**.

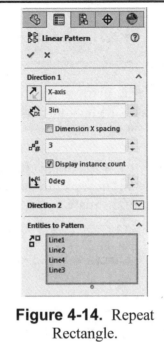

Figure 4-14. Repeat Rectangle.

Figure 4-15. Extrude Menu.

Now save your model. Pull down the **File** menu and select **Save As**. On the "Save As" menu, select your appropriate file folder, type in the part name **DRAWER TRAY.SLDPRT**, then click **Save**.

To finish this exercise, you should print a hard copy for submission to your instructor. Open your **TitleBlock-Inches.drwdot**. Save your drawing sheet as **DRAWER TRAY.slddrw**. Insert the image into the **Title Block** drawing sheet like in previous labs. **Print** a hard copy to submit to your lab instructor. The drawing should look like **Figure 4-16**.

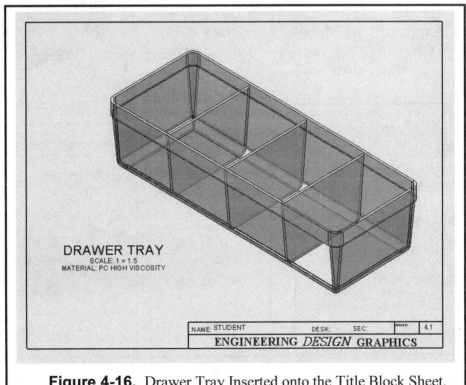

Figure 4-16. Drawer Tray Inserted onto the Title Block Sheet.

Exercise 4.2: TAP-LIGHT DOME

Go to **File** and open **ANSI-INCHES.prtdot**. Pull down **File** again, select **Save As**, then name it **TAP-LIGHT DOME.sldprt** and save the file in your folder.

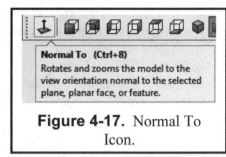

Figure 4-17. Normal To Icon.

Select the **Top Plane** in which to work, then click on the **Normal To** icon which looks like a square plate with an arrow pointing upwards out of the center of the plate. (See **Figure 4-17**.)

Select the **Sketch** icon and draw a **Circle** at the **ORIGIN** with a <u>**Diameter**</u> of **4.70** inches.

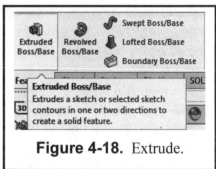

Figure 4-18. Extrude.

Select the **Extrude** symbol (**Figure 4-18**) on the left side of the screen. Do a **Blind Extrude** of **0.375** inches. Click on the **OK** button, which will produce an image similar to **Figure 4-19**.

Click on the words "**Extrude**" in the Feature Manager twice, (this is not a double click) and rename this feature as "**Dome Base**."

Select the **Top** plane of the **Dome Base**. Now go to the **Insert** tab **Select - Features** and **Select - Dome**. In the "**Dome**" features window give the **Dome Height** of **1.20 inches** (See **Figure 4-20**) and click **OK**. The dome operation should give you an object similar to **Figure 4-21**.

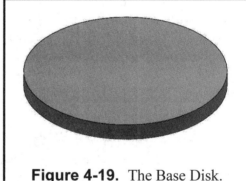

Figure 4-19. The Base Disk.

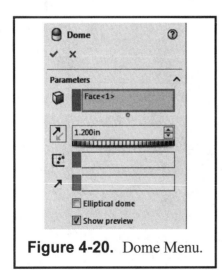

Figure 4-20. Dome Menu.

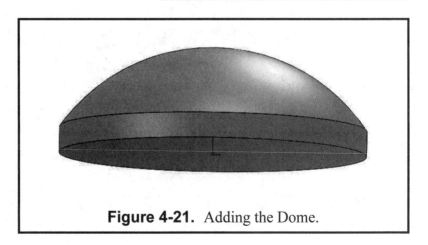

Figure 4-21. Adding the Dome.

Rotate the model so you can see the bottom face. Select the bottom face and pick the **Shell** icon (**Figure 4-22**) in the Features Tool Bar. Give the shell thickness of **0.075** in the parameters box and **OK** the selection.

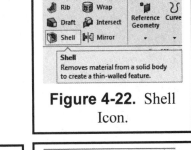

Figure 4-22. Shell Icon.

You might have to zoom in to select the bottom thin surface of the resulting shell. After you have selected the thin surface, activate the **Sketch Tab, Select** the **Sketch Icon** and then select the **Convert Entities** icon (**Figure 4-23**). This converts the outer edges of that surface into a new sketch entity that can be used for a new feature. Next, select the **Offset Entities** icon (**Figure 4-24**) and pick the black edge of the converted entity; enter the distance of **0.20** (**Figure 4-25**) outward so the new offset pattern lies outside of the Dome (**Figure 4-26**). **Extrude** this pattern upward by **0.075** inches.

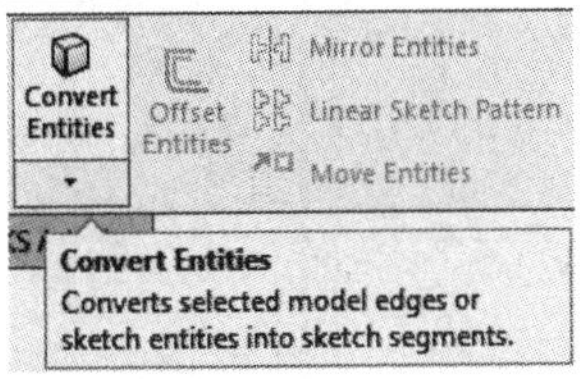

Figure 4-23. Convert Entities.

Figure 4-24. Offset Icon.

This should complete the basic shelled dome. The model should resemble **Figure 4-27**. What remains to be done is to cut the four holes in the lip of the dome where fastening screws pass to hold the body and the lens frame together. The holes in the lip of the dome are placed at rather unusual angles. This is done to ensure that the dome can only be placed on the body of the light fixture in one position.

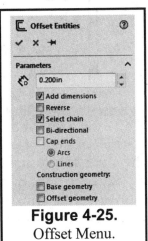

Figure 4-25. Offset Menu.

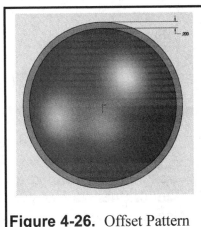

Figure 4-26. Offset Pattern

Click on the top surface of the outer lip of the domed lens and select the **Normal To** icon. Click on the **Sketch** icon and draw a vertical **Centerline** through the origin. Next, **Draw four non-constrained radiating centerlines (no Blue Dotted lines attached to the end points)** using the centerline option and a **Bolt circle with a diameter of 5.05** inches (be sure to check **FOR CONSTRUCTION** in the option box) as shown in **Figure 4-28**. The large bolt circle must be drawn since it is not the same diameter as the outer rim of the dome lip. Dimension the four radiating centerlines as shown in **Figure 4-29**.

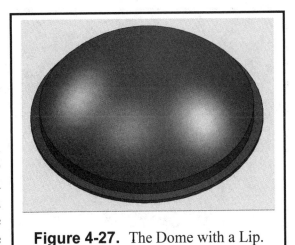

Figure 4-27. The Dome with a Lip.

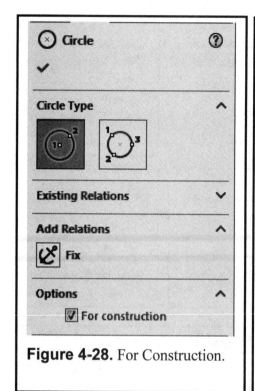

Figure 4-28. For Construction.

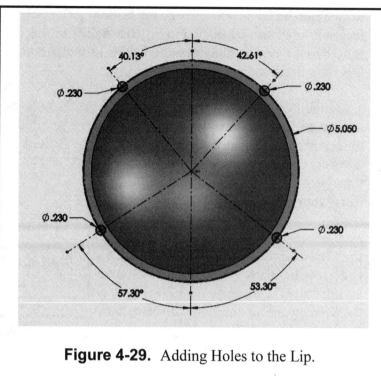

Figure 4-29. Adding Holes to the Lip.

Locate the four small holes at the intersections of the radial lines and the construction circle by using **Quick Snaps – Intersection Snap**. To get this menu, select the **Sketch – Circle** command, then **right click the mouse in an open area of the drawing screen and Select** the **Quick Snaps** option **(See Figure 4-30)**. Dimension the four circles and do the **Extruded Cut, Through All**.

Your final model should look similar to **Figure 4-31**.

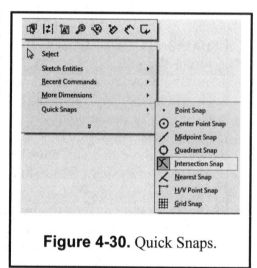

Figure 4-30. Quick Snaps.

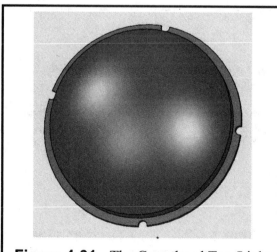

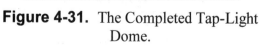

Figure 4-31. The Completed Tap-Light Dome.

Right click on **Material (not specified)** in the Feature Manager. Select **Edit Material**. Expand the **Plastics Tab** and select **ABS** and then select **Apply** and **Close**. Now save your model. Pull down the **File** menu and select **Save As**. On the "Save As" menu, select your appropriate file folder, type in the part name **TAP-LIGHT DOME.sldprt** and then click **Save**.

To finish this exercise, you should print a hard copy for submission to your instructor. First open your **TitleBlock-Inches.drwdot**. Follow the instructions given in Unit 1 for inserting the rendered image onto a Title Block.

Save your drawing as **TAP-LIGHT DOME.slddrw**. Your final drawing should look similar to **Figure 4-32**.

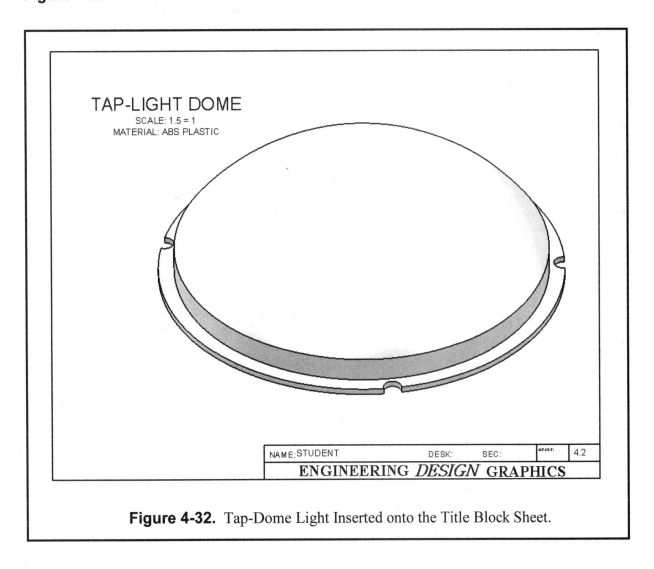

TAP-LIGHT DOME
SCALE: 1.5 = 1
MATERIAL: ABS PLASTIC

| NAME: STUDENT | | DESK: | SEC: | QRSGC: | 4.2 |

ENGINEERING *DESIGN* GRAPHICS

Figure 4-32. Tap-Dome Light Inserted onto the Title Block Sheet.

Exercise 4.3:
THREADS AND FASTENERS

THREAD TYPES

SQUARE THREADS & ACME THREADS These Threads have been used for many years to transmit power either from a turning motion or a linear motion. The uses have increased as more and more processes require this type of translation. Since the square thread is difficult to manufacture, the Acme thread has become the more commonly used thread for the purpose of translating power. **Figure 4-33** shows the profile of an Acme Thread and **Figure 4-34** shows a Square Thread profile.

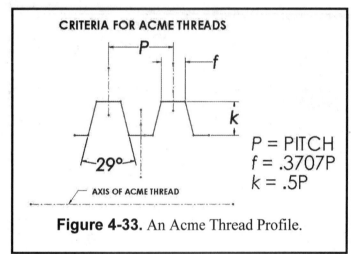

P = PITCH
f = .3707P
k = .5P

Figure 4-33. An Acme Thread Profile.

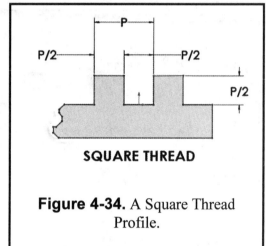

Figure 4-34. A Square Thread Profile.

BUTTRESS THREAD This thread form is a nonsymmetrical thread that is used where exceptionally high stresses lie along the axis of the threaded shaft. An example of this thread form is shown in **Figure 4-35**.

STANDARD V-THREADS - This thread form is the basis for most nuts and bolts; however, within this thread form there are many deviations that are used for special applications. Several of the most common standards are the American Standards and the Metric

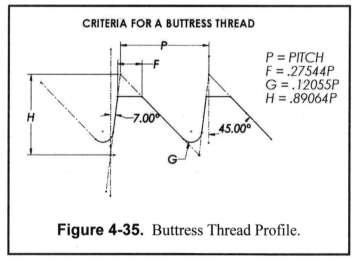

P = PITCH
F = .27544P
G = .12055P
H = .89064P

Figure 4-35. Buttress Thread Profile.

Standards. Within the American standards the most common series are "UNC" or Unified National Course (also known as "NC" National Course); "UNF" or Unified National Fine (also known as "NF" National Fine); and "UNEF" or Unified National Extra Fine (also known as "NEF" National Extra Fine). Within the Metric standards there is a coarse series and a fine series that are distinguished only by the different values of the pitch.

THREAD TERMINOLOGY

Following is a list of terms that are used extensively in association with threads:

> ➤ Major Diameter – the largest diameter of an internal or external thread.
> ➤ Pitch – The distance from one crest of a thread to the next crest.
> ➤ Pitch Diameter – the theoretical diameter at the point where the tooth width and gap width are equal.
> ➤ Minor Diameter - the smallest diameter of an internal or external thread.
> ➤ Threads per Inch – A number of threads in an inch that determines the pitch.
> ➤ Thread Depth – generally the difference between the major diameter and minor diameter divided by 2.

THREAD NOTES

Thread notes are the most critical portion of a thread representation. It gives all of the information that is necessary for a machinist to produce the thread or the manufacturer to select the correct fasteners for assembly. In **Figure 4-36** an American thread note is illustrated and in **Figure 4-37** is a metric thread note.

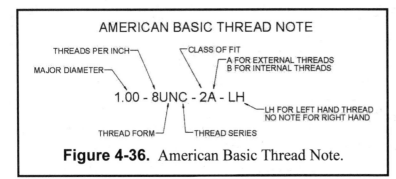

Figure 4-36. American Basic Thread Note.

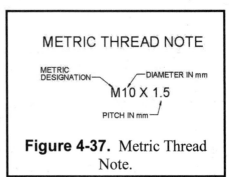

Figure 4-37. Metric Thread Note.

The dimensions of threads are based primarily on the number of threads per inch in the American system. The number of threads divided into one inch produces the pitch, which is also the basis in the metric system. In the exercise we will be using the criteria for internal and external threads. **Figure 4-38** shows the criteria required for external threads and **Figure 4-39** gives the specifications for the internal thread. The values for the different formulas are given in the various tables provided in Machinist or Engineering Handbooks. **Figure 4-40** is an example of such a table and includes the criteria that will be used in **Exercise 4.3**.

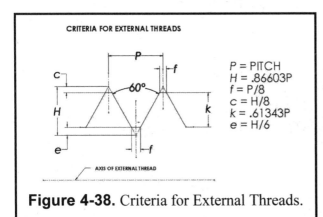

Figure 4-38. Criteria for External Threads.

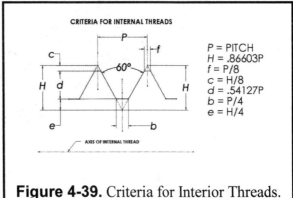

Figure 4-39. Criteria for Interior Threads.

| UNIFIED COURSE THREAD SERIES PARTIAL TABLE | | | | | |
Fractional Sizes	Basic Major Diameter	Threads per Inch	Pitch	Minor Diameter Internal Threads	Thread Depth
1/4	0.250	20	0.0500000	0.1959	0.0325
5/16	0.313	18	0.0555555	0.2524	0.0361
3/8	0.375	16	0.0625000	0.3073	0.0406
7/16	0.438	14	0.0714286	0.3602	0.0464
1/2	0.500	13	0.0769231	0.4167	0.5000
9/16	0.563	12	0.0833333	0.4723	0.0541
5/8	0.625	11	0.0909091	0.5266	0.0590
3/4	0.750	10	0.1000000	0.6417	0.0650
7/8	0.875	9	0.1111111	0.7547	0.0722
1	1.000	8	0.1250000	0.8647	0.0767
1 - 1/8	1.125	7	0.1428571	0.9704	0.0928
1 - 1/4	1.250	7	0.1428571	1.0954	0.0928
1 - 1/2	1.500	6	0.1666667	1.3196	0.1083

Figure 4-40. Unified Course Thread Table for Selected Sizes.

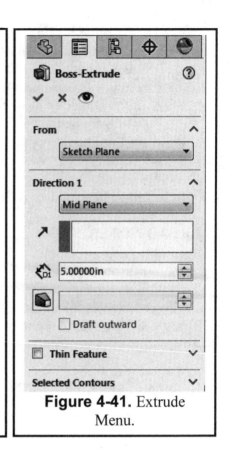

Figure 4-41. Extrude Menu.

Begin in the usual manner by going to your folder and opening **ANSI-INCHES.prtdot**. Pull down **File** again, select **Save As**, then name it **SHAFT THREAD.sldprt** and save the file in your folder. Go to **Tools – Options** and change the units to **5 Decimal Places**.

Select the **Right Plane** to begin your drawing. **Sketch** a **Circle** at the **Origin** that has a **Diameter** of **1.00** inch and **Extrude** it **Mid Plane** to **5** inches. The result will look like **Figure 4-42**.

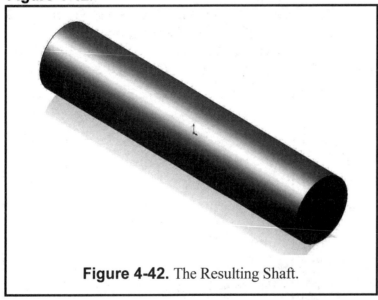

Figure 4-42. The Resulting Shaft.

For the sake of proper clearances between interior and exterior threads, the exterior thread for a 1 inch thread is undersized by a tolerance of between 0.9966 and 0.9744. In our instance we will use the mid-value of 0.9855. Select the left end of the shaft, **Select Sketch** and **Draw** a circle with a diameter of **0.9855**. On the **Features** Tab, select the **Extruded Cut - Blind** option. Check the **Flip Side to Cut** box for a distance of **3.25** inches; see **Figure 4-43**.

Now we will cut threads on this shaft for **3.125** inches from the left end. **Select** the surface of the **Left End** of the shaft and select the **Features – Chamfer** option and using the **Thread Depth** formula in **Figure 4-38 on page 4-11**: (.61343*.125) or **.0767**.

Next, select the **Left** end of the shaft and select the **Insert - Reference Geometry - Plane** and enter the **Pitch** distance for a 1" thread from the Table in **Figure 4-40**. The chamfer and the inserted plane should look like **Figure 4-44.**

Select **Plane 1**, or the plane you just created. Then activate the Sketch tab and Select **Sketch**. Select the large circle of the chamfer and select **Convert Entities**. Now go to the **Insert** tab, pick the **Curve** option and select **Helix/Spiral**. In the box in the Feature Manager tree, supply the following information.
Defined By: **Height and Pitch**
Parameters: **Constant Pitch**

    Height: **3.25**
    Pitch: **0.125**
Check "**Reverse Direction**"
    Start Angle: **270 degrees**
Check: **Clockwise**
Okay the criteria.

Because we placed the initial construction at the origin and placed the start angle of the helix at 270 degrees, it is now possible to draw the tooth profile on the front plane. In the Feature Manager Tree, select the **Front Plane**. Go to the Front View, go to the **Sketch Tab** and **Select Sketch**. Zoom into the left end of the shaft as shown in Figure 4-45. **Draw a centerline horizontally** along the bottom edge of the shaft. **Draw a second centerline vertically** along the edge view of Plane 1. With the vertical centerline active, go to **Tools – Sketch Tools – Dynamic Mirror**. This puts two small dashes across the ends of the vertical centerline to indicate that the dynamic mirror function is active. See **Figure 4-45**.

<u>Tooth Profile</u>. **Sketch** a half profile of the tooth only on the right side of the centerline. The **Dynamic Mirror** command will produce the other half of the profile. **Dimension** your sketch to correspond to those shown in **Figure 4-46**. Make sure the wide base of the tooth profile extends very slightly below the horizontal centerline. The dimension of the small end of the tooth profile is P/8. Since P = .125, .125/8 = **0.0156**. The dimension from the horizontal centerline to the small end

Figure 4-43. Flip Side Cut.

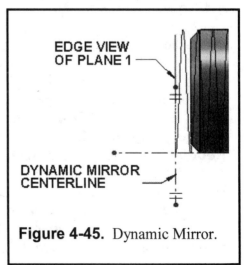

Figure 4-44. Chamfer and Inserted Plane.

EDGE VIEW
OF PLANE 1

DYNAMIC MIRROR
CENTERLINE

Figure 4-45. Dynamic Mirror.

is 0.0767 obtained from the formula in **Figure 4-38** that shows that dimension is 0.61343P or 0.61343 * .125 = **.076678**.

Click **Rebuild** to get out of all pending operations. Go to a pictorial view to show how the **Swept Cut** operation will affect your threaded shaft. An isometric view should show your profile and sweep path similar to **Figure 4-47**.

Swept Cut. Now you are ready to create the completed thread. Activate the Features tab and select the **Swept Cut** icon and observe the **Swept Cut** menu that appears. Under **Profile and Path**, using the mouse, first select the Tooth Profile and then the Helix/Spiral. Your completed thread should appear in dimetric view similar to **Figure 4-48**.

Now **Select** the **Right End** of the shaft and select the **Sketch Icon** then draw a **Hexagon** at the origin that is 1.5 inches across the flat edges of the hexagon. Draw the cursor horizontally to place the right corner so it constrained to the right of the origin. Next, use the smart dimension to fix the distance from the top of the hexagon to the bottom to be **1.50** inches. This completes the sketch. Now **Extrude** the sketch away from the threaded end of the bolt to a depth of **39/64"** or **0.609375"**. This measurement is derived from the Standards for a 1" Thread. See **Figure 4-49** to view this operation.

Select the visible face of the Hexagon to begin your next sketch. Activate the Sketch Tab and select the Sketch Icon. Draw a Circle at the origin and make it tangent to one of the flat edges of the Hexagon, **Figure 4-50**. Select the **Rebuild Icon** and then **Select** the **Top Plane** then go to the top view to begin the next sketch. After you select the **Sketch Icon**, **Select** the **Convert Icon**. **Select the previous sketch** to convert. After you click on the Green Check Mark,

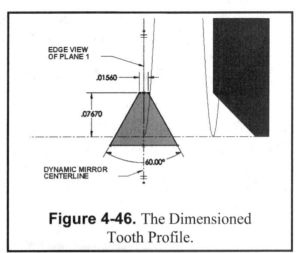

Figure 4-46. The Dimensioned Tooth Profile.

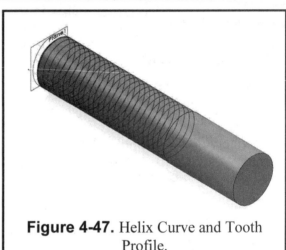

Figure 4-47. Helix Curve and Tooth Profile.

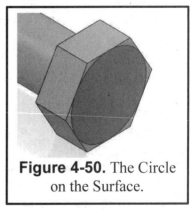

Figure 4-48. Completed Thread.

Figure 4-49. The Extruded Profile.

Figure 4-50. The Circle on the Surface.

Select the line that was generated and **select "For Construction"** in the **Feature Manager Tree**. All of this is required to actually begin the sketch for the next feature. **Select** the **Line Command** and begin the Sketch at the upper point of the construction line. Draw a line at an angle across the corner of the nut then horizontally to the vertical surface of the nut and then back down to the original point of the sketch. Dimension the sketch angle to equal to 30 degrees as shown in **Figure 4-51**. Also draw a Center line horizontally through the origin. This completes the sketch. Select the Features Tab and **Revolve Cut** the 30 degree triangle about the centerline. The 1" bolt is now completed as shown in **Figure 52**.

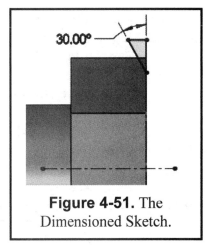

Figure 4-51. The Dimensioned Sketch.

CONSTRUCTION OF A CORRESPONDING NUT

There are many standards dictating the production of threads and fasteners. When beginning to construct fasteners it is advisable to consult official standards to be sure of your correctness. In the case of the 1" diameter nut that would fit the above threaded shaft, the hexagon of the nut is 1.5 times the major diameter of the shaft. Begin in the usual manner by going to your folder and opening **ANSI-INCHES.prtdot**. Pull down **File** again, select **Save As**, then name it **THREADED NUT.sldprt** and save the file in your folder. Go to **Tools – Options** and change the units to **5 Decimal Places**.

Now **Select** the **Right Plane** and select the **Sketch Icon** then draw a **Hexagon** at the origin that is 1.5 inches across the flat edges of the hexagon. Draw the cursor horizontally to place the right corner so it constrained to the right of the origin. Next, use the smart dimension to fix the distance from the top of the hexagon to the bottom to be **1.50** inches. Now referring to **Figure 4-40**, **Page 4-12** you will draw a **Circle** at the origin to be equal to the Minor Diameter for Internal Threads which is **0.8647** inches. This completes the sketch. Now **Extrude** the sketch at **Mid-Plane** to a depth of **55/64"** or **0.859375"**. This measurement is derived from the Standards for a 1" Thread. See **Figure 4-54** to view this operation.

Select the visible face of the Hexagon to begin your next sketch. Activate the Sketch Tab and select the Sketch Icon. Draw a Circle at the origin and make it tangent to one of the flat edges of the Hexagon, **Figure 4-55**.

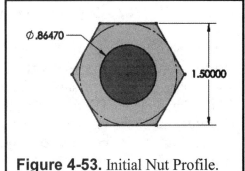

Figure 4-52. The Completed Bolt

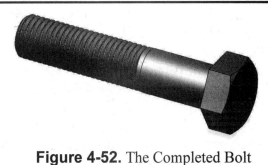

Figure 4-53. Initial Nut Profile.

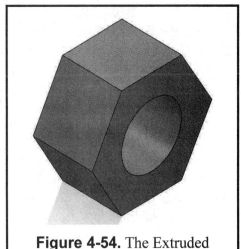

Figure 4-54. The Extruded Profile.

Select the **Rebuild Icon** and then **Select** the **Top Plane**, then go to the top view to begin the next sketch. After you select the **Sketch Icon, Select** the **Convert Icon. Select the previous sketch** to convert. After you click on the Green Check Mark, **Select** the line that was generated and **select** "For Construction" in the **Feature Manager Tree**. All of this is required to actually begin the sketch for the next feature. **Select** the **Line Command** and begin the Sketch at the upper point of the construction line. Draw a line at an angle across the corner of the nut then horizontally to the vertical surface of the nut and then back down to the original point of the sketch. Dimension the sketch angle to equal to 30 degrees as shown in **Figure 4-56**. Also draw a Center line horizontally through the origin. This completes the sketch. Select the Features Tab and **Revolve Cut** the 30 degree triangle about the centerline. Now **Select** the **Mirror Feature** and select the **Right Plane**. This will repeat the Revolved Cut on the other side of the nut. **See Figure 4-58**.

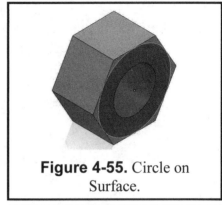

Figure 4-55. Circle on Surface.

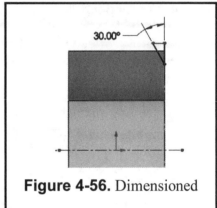

Figure 4-56. Dimensioned

The next operation is to construct a **Chamfer** inside both ends of the hole, equal to the Thread Depth which is in the Chart found in **Figure 4-45**, **Page 4-13** for the 1 inch Diameter, (**0.0767**). Select the Green Check Mark to accept the operation. **See Figure 4-57.**

Select the visible end of the nut. Then go to **Insert – Reference Geometry – Plane**. Place it 1 Pitch distance away from the selected surface. Pitch distance is **0.125"**. With Plane 1 activated, select the **Sketch Icon** and **Convert** the small diameter. Now select **Insert – Curve – Helix/Spiral.** In the **Defined By: Select Height and Pitch.** For the Height enter something a little larger than 1" and the **Pitch** is 8 per inch or **.125"**. The **Start Angle** should be at **270 Degrees**. Okay these settings, then **Select** the **Front Plane** and go to the front view. **Select** the **Sketch Icon** and draw **Two Centerlines**. The first

Figure 4-57. Revolved Cut.

one should be horizontal and coincident at the bottom of the Hole. Turn off the shading and turn on the hidden lines to see where to place this line. The second centerline is vertical along the edge view of **Plane 1**. While this line is active, go to **Tools – Sketch Tools – Dynamic Mirror**. This puts small horizontal dashes on either end of the last centerline you drew. Draw the tooth profile starting on the centerline and the Pattern will be duplicated across the centerline. Dimension the tooth profile as shown in **Figure 4-59**. These dimensions are derived from the formulas in **Figure 4-38**. Rebuild your solid model and go to the **Features Tab** and **Select Swept Cut**. First select the tooth profile, then the Helix to cut the thread into the nut. Your final part should look like **Figure 4-60**.

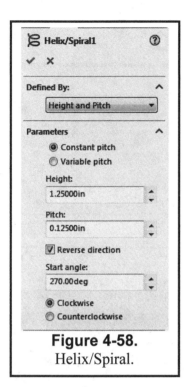

Figure 4-58.
Helix/Spiral.

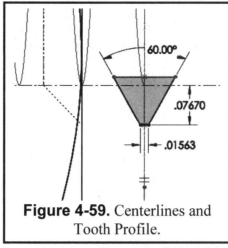

Figure 4-59. Centerlines and Tooth Profile.

Figure 60. The Finished Nut.

Save your part as **Threaded Nut.SLDPRT**.

To finish this exercise, you should print a hard copy for submission to your instructor. First open your **TitleBlock-Inches.drwdot**. Insert the two images into the **Title Block** drawing sheet as in **Figure 4-61**. Save your drawing as **SCREW THREADS.slddrw**. **Print** a hard copy to submit to your lab instructor.

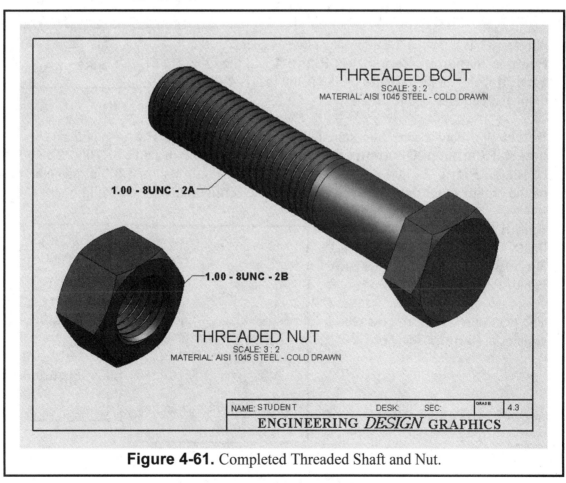

Figure 4-61. Completed Threaded Shaft and Nut.

Exercise 4.4: JACK STAND

Go to **File** and open **ANSI-INCHES.prtdot**. Pull down **File** again, select **Save As**, then name it **JACK STAND.sldprt** and save the file in your folder.

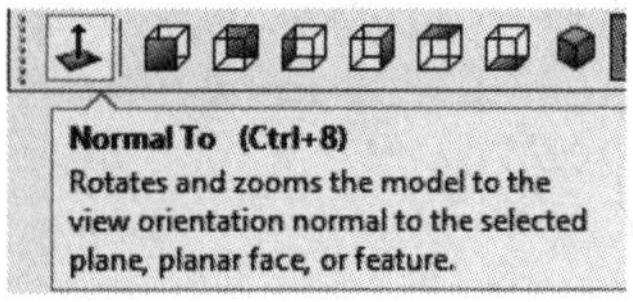

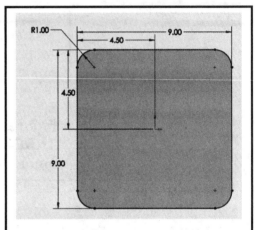

Figure 4-62. Normal To Icon.

Select the **Top Plane** in which to work, then click on the **Normal To** icon which looks like a square plate with an arrow pointing upwards out of the center of the plate (**Figure 4-62**).

Select the **Sketch** icon and draw a **Rectangle** centered on the origin. Constrain the rectangle as shown in **Figure 4-63**. Add **1" Fillets** to the corners. **Select** the rebuild icon (the stop light).

Before you can proceed with the next step you will have to establish a new work plane. To better see what is going to happen, go to the isometric view. Select the **Top Plane** and then select **Insert, Reference Geometry, Plane**. Enter the distance of **6.5"**.

After you have created the reference plane, select it in the **Feature Manager Tree**. Select **Plane 1** and then the **Sketch** icon. Draw a **Circle** on the plane that is **4.0"** in diameter, centered on the origin.

Figure 4-63. Rectangle Sketch.

Select the rebuild icon (the stop light). **Select** the **Top Plane** again and then select **Insert, Reference Geometry - Plane** again. Enter the distance of **10"**. This plane will be labeled **Plane 2**. Select it and **Sketch** a **Circle** on it that is **3.0"** in diameter, centered on the origin. Go to an isometric view to see the configuration as shown in **Figure 4-64**.

Select the rebuild icon (the stop light). **Select** the **Lofted Boss/Base** icon in Features menu. (See **Figure 4-65**.)

The next operations are best seen in a pictorial (**Isometric**) view.

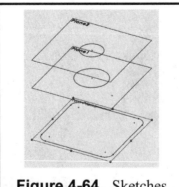

Figure 4-64. Sketches for Lofting.

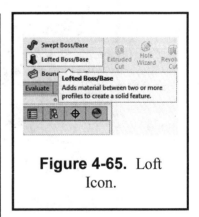

Figure 4-65. Loft Icon.

Select the three sketches in the order that they are to be lofted. The **bottom square first**, then the **circle in the middle**, and last, the **top circle**. The order of the sketch selection will appear in the **Base – Loft Menu** (See **Figure 4-66)**. As you select the sketches a preview of the loft will be shown on the screen. If it appears as the figure in **Figure 4-67**, select the (√) to finalize the process.

Rotate the model so you can see the **bottom face**. **Select** the **bottom face** and pick the **Shell** icon (**Figure 4-68**) in the features tool bar. Give the shell thickness of **0.1875"** in the parameters box and (√) **OK** the selection.

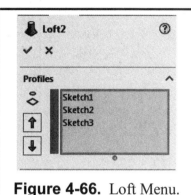

Figure 4-66. Loft Menu.

Figure 4-67. Lofted Part.

Several other operations must be performed in order to make this a viable design. The top must be reinforced for a threaded adjustment screw and the sides are to be hollowed out to lighten the weight of the jack stand. (The thread formation has been omitted from this design).

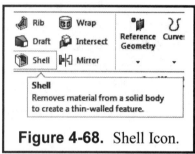

Figure 4-68. Shell Icon.

Select the **Top Surface** of the jack stand and **Select** the **Normal-To** icon. Select the Sketch tab and Select **Sketch,** then **Convert** the entity and do a **Boss Extrude** downward for **3"**. **Select** the **Top Surface** again and **Sketch** a **1.25** diameter **Circle**. Do an **Extruded Cut, Through all** to create the hole as shown in **Figure 4-69**.

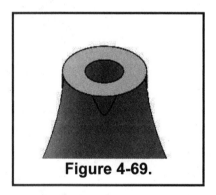

Figure 4-69.

The last two operations are the openings that need to be cut in the sides of the jack stand. **Select** the **Front Plane** and **Select**

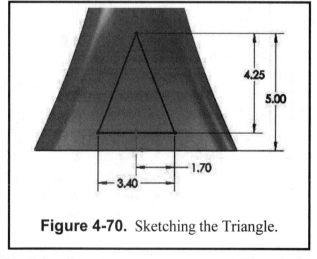

Figure 4-70. Sketching the Triangle.

Normal-To. Sketch the **Triangle** and set the dimension constraints as shown in **Figure 4-70**. Also add a **vertical relation** between the top point of the triangle and the origin. **Execute** a **Cut Extrude, Through All** in **Both Directions**.

Since you will be using the exact same sketch in the Right Plane, click on the **Previous Sketch,** go to the **Windows Edit** menu and select **Copy** (see **Figure 4-71**). Next, go to the **Feature Manager Tree**, select the **Right Plane**, and **Go** to the **Windows Edit** menu and select **Paste**. Select the **Sketch** that resulted from this operation and repeat the **Cut Extrude, Through All** in **Both Directions** to complete the model. Your final model should look similar to **Figure 4-72**.

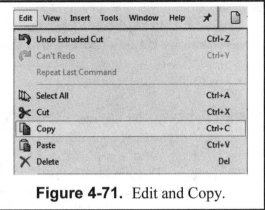

Figure 4-71. Edit and Copy.

Now save your model. Pull down the **File** menu and select **Save As**. On the "Save As" menu, select your appropriate file folder, type in the part name **JACK STAND.sldprt**, then click **Save**.

To finish this exercise, you should print a hard copy for submission to your instructor. First open your **TitleBlock-Inches.drwdot**. Follow the instructions given in Unit 1 for inserting the rendered image onto a Title Block. Save your drawing as **JACK STAND.slddrw** in your designated folder.

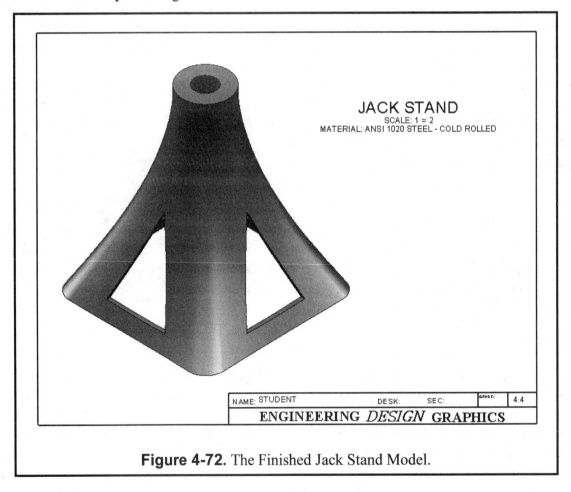

Figure 4-72. The Finished Jack Stand Model.

SUPPLEMENTARY EXERCISE 4-5: TAPE DISPENSER

Using the **ANSI-INCHES.prtdot**, draw the profile of the Tape Dispenser and extrude it 0.75 inches. Select the front surface and fillet it to 0.05 inches. Select the back surface and the two interior surfaces as indicated in the second figure and SHELL the solid to a thickness of 0.10 inches. Open your **TitleBlock-Inches.drwdot**. Follow the instructions given in Unit 1 for inserting the rendered image onto a Title Block. Save your drawing as **TAPE DISPENSER.slddrw** in your designated folder.

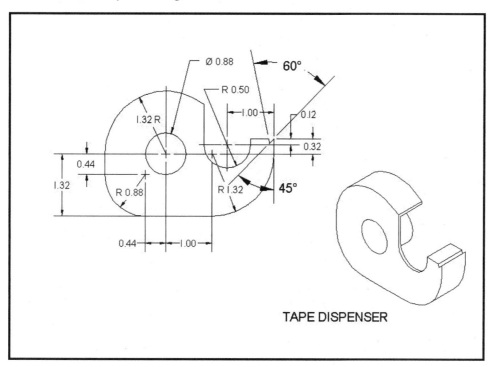

TAPE DISPENSER

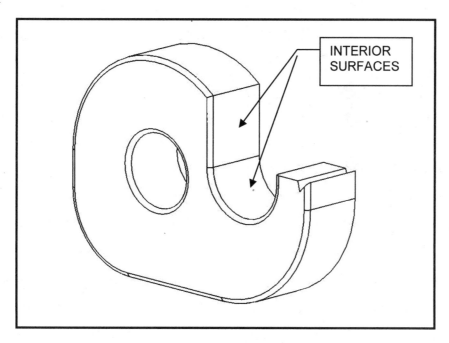

INTERIOR SURFACES

SUPPLEMENTARY EXERCISE 4-6: FUNNEL

Build a full size model of the Funnel below. Begin in the **Top Plane**. To get the desired model, each section must be lofted separately. Activate the right plane and sketch a guide curve connecting the left hand edges of the first two sketches. Repeat the process for the other two lofts. When the lofting process is completed, select the top and the bottom surfaces and shell the model to a wall thickness of 0.10 inches. Insert it onto a Title Block and title it **FUNNEL.slddrw**.

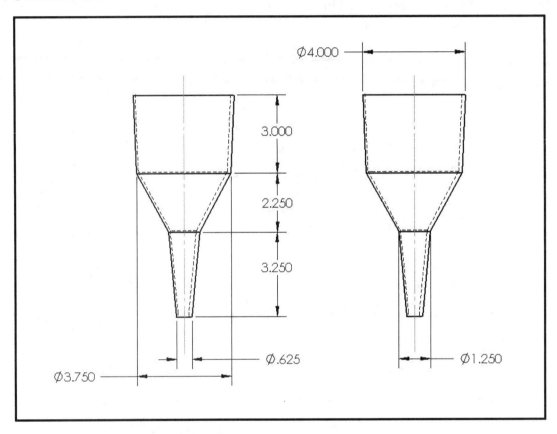

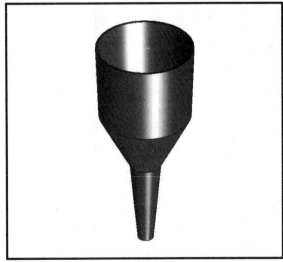

SUPPLEMENTARY EXERCISE 4-7:
TWO INCH THREAD – UNC

1. Construct a 2" diameter shaft 5 inches long.
2. The major screw thread diameter is from 1.9868 – 1.9424. Cut the shaft down to 1.9646 for three inches.
3. Screw thread note - 2-4 ACME-2A X 3.
4. Construct a chamfer equal to the Thread Depth (the "k" value in the Definition of Terms) on the end of the shaft that will be threaded.
5. Insert a reference plane one pitch "P" distance away from the end of the shaft.
6. Select the reference plane and convert the 1.9646 diameter circle.
7. Insert – curve – helix spiral with the parameters – height and pitch. The height will be 3.25 and the pitch will be .25, with the starting angle being at 270 degrees – clockwise.
8. Select the Front plane and draw a centerline along the lower edge of the shaft and a vertical centerline through the edge of the inserted plane. Select the vertical Centerline and use the Dynamic Mirror function.
9. Sketch the tooth profile according to the given parameters.
10. Use the Features – Swept Cut function to cut the thread.

ACME THREAD SPECIFICATIONS					
Nomial Size	Basic Major Diameter	Threads per Inch	Pitch	Thread Depth d = .5P + 0.01	Width of Space at Bottom of Thread W = .3707P - .0052
2	1.9646	4	0.25	0.135	0.087475

Values from Thread Table

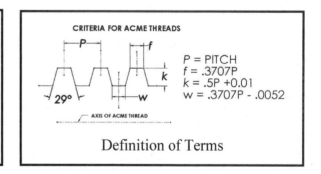

$P = PITCH$
$f = .3707P$
$k = .5P +0.01$
$w = .3707P - .0052$

Definition of Terms

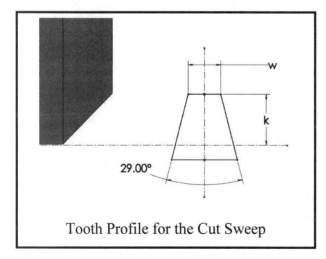

Tooth Profile for the Cut Sweep

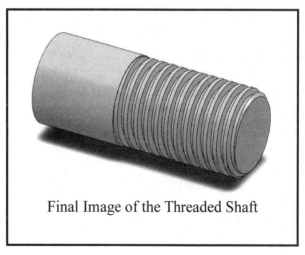

Final Image of the Threaded Shaft

NOTES:

Computer Graphics Lab 5: Assembly Modeling and Mating

In Computer Graphics Lab 5, you will build several parts of an assembly. You will then assemble the parts and fit them properly together using mating functions available in SOLIDWORKS. This introductory section will get you started and then two exercises, the Terminal Support Assembly and the Swivel Eye Block Assembly, are provided.

ASSEMBLY FILE

In the previous exercises, you started with a **File, New, Part** command sequence to start a new part file. SOLIDWORKS offers two other choices when starting a new file, **Assembly** and **Drawing**, as shown in **Figure 5-1**. This is the normal way to start an Assembly.

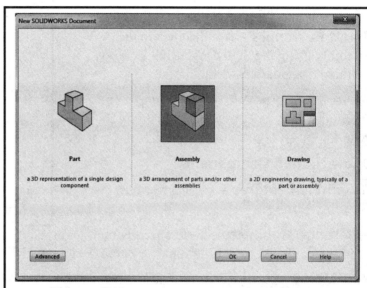

Figure 5-1. Starting a New Assembly File.

TILING THE WINDOWS

There are different ways to start the assembly. One easy way is to open all the parts to the assembly in SOLIDWORKS, in addition to having the new assembly file open. You can then **Tile Vertically** or **Horizontally** the windows (see **Figure 5-2**) and then simply "Drag and Drop" the parts into the assembly file. The first part placed into the assembly is fixed and cannot be moved. The others can be fixed in relation to the first part by using the Mate commands.

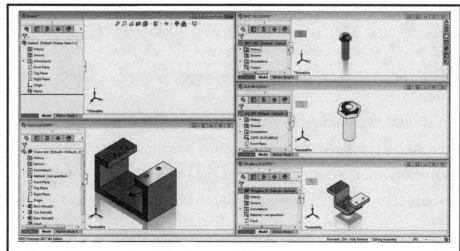

Figure 5-2. Tiling the Windows Vertically to Begin the Assembly.

THE ASSEMBLY TOOLBAR

The Assembly toolbar is located on the left hand side of the screen when you are in the assembly mode. It contains the following commands:

Insert Components – Adds an existing part or sub-assembly to the assembly.

Mate – Positions two components relative to one another.

Linear Component Pattern – Patterns components in one or two linear directions.

Smart Fasteners – Add features to the assembly using the SOLIDWORKS Toolbox Library of standard hardware.

Move Component – Moves a component within the degrees of freedom defined by its mates.

Show Hidden Component

Assembly Features – Creates various assembly features.

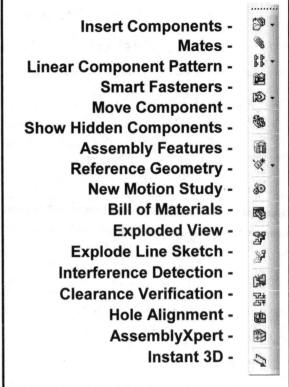

Insert Components -
Mates -
Linear Component Pattern -
Smart Fasteners -
Move Component -
Show Hidden Components -
Assembly Features -
Reference Geometry -
New Motion Study -
Bill of Materials -
Exploded View -
Explode Line Sketch -
Interference Detection -
Clearance Verification -
Hole Alignment -
AssemblyXpert -
Instant 3D -

Figure 5-3. The Assembly Toolbar.

Reference Geometry – Reference geometry commands: Plane, Axis, Coordinate System, Point, and Mate Reference.

New Motion Study – Inserts a new motion study.

Bill of Materials

Exploded View – This separates the components into an exploded view.

Explode Line Sketch – This can be used to add or edit a 3-D line sketch, showing the relationship between exploded components.

Interference Detection – This can be used to detect any interference between components in an assembly.

Clearance Verification – Verifies clearance between components.

Hole Alignment – This checks assembly hole alignment.

AssemblyXpert – Displays statistics and checks the health of the current assembly.

Instant 3D - This enables dragging of handles, dimensions, and sketches to dynamically modify features.

MATE TYPES

You can mate parts in an assembly by clicking the entities in the computer space and then select the **Mate** icon. The available list of mates is then displayed and listed below. *Note*: Only those mate types that are valid for the selected entities are available when this command is used. The valid mating relationships are:

Coincident - positions selected faces, edges, and planes (in combination with each other or combined with a single vertex) so they share the same infinite line. Coincident positions two vertices so they share a common point.

Perpendicular - places the selected items at a 90-degree angle to each other.

Tangent - places the selected items in a tangent mate (at least one selection must be a cylindrical, conical, or spherical face).

Concentric - places the selections so that they share the same center point.

Parallel - places the selected items so they lie in the same direction and remain a constant distance apart from each other.

Distance - places the selected items with the specified distance between them.

Angle - places the selected items at the specified angle to each other.

Symmetry - places the selected items at an equal distance from a plane of symmetry.

VIEWING ASSEMBLIES

When you are building assemblies with many components, sometimes viewing of specific parts becomes difficult. One way to view a part in an assembly is to set the transparency of the other parts. Select the **Tools**, **Options**, **Display/Selection** tab. Then set the Assembly Transparency to **Force Assembly Transparency** with the slider at some value up to 100%. An example is shown in **Figure 5-4**.

EXPLODING ASSEMBLIES

Assembly components can be exploded using the **Exploded View** icon on the assembly toolbar. This produces the **Assembly Exploder** menu on the screen, as shown in **Figure 5-5**. More on exploding assemblies will be covered in **Unit 8**.

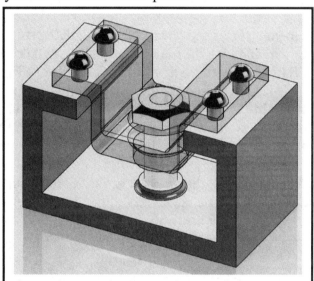

Figure 5-4. Setting the Transparency of Parts in an Assembly.

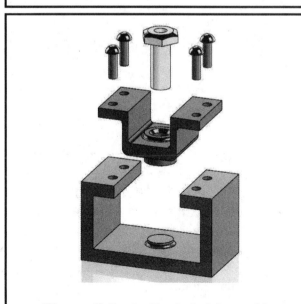

Figure 5-5. An Exploded Assembly.

Exercise 5.1: TERMINAL SUPPORT ASSEMBLY

The Terminal Support Assembly has three major components: the Frame, the Wing Base and the Pin. It also contains four standard rivet parts for attachment to the frame. You will start by designing the Frame part. After that you will model the Wing Base, the Pin and then finally create one version of the rivet. You will mate all of them in an assembly file, including bringing in the rivet four times.

FRAME

Go to your folder and open **ANSI-INCHES.prtdot**. A good practice is to immediately name your part and save the file in your folder. Pull down **File** again, select **Save As**, and then name it **FRAME.sldprt**. Next, click the **LMB** on the **Front** plane icon in the Feature Manager tree and also click the **Front** view orientation. Enter the **Sketch** mode and draw a vertical **Centerline** through the origin. Then go to **Tools, Sketch Tools**, and select **Dynamic Mirror**. Now, draw the right half of the image in **Figure 5-6**. You will see the left half drawn at the same time. When finished with the sketch, extrude it **4.00 inches – Mid Plane**.

Select the **TOP** of the recessed surface of the resulting Frame and enter the Sketch Mode. Place the four holes in the sketch according to the dimensions given in **Figure 5-7**. Perform an **Extrude Cut – Blind** only far enough to penetrate the upper lips of the Frame.

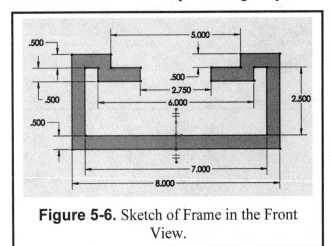

Figure 5-6. Sketch of Frame in the Front View.

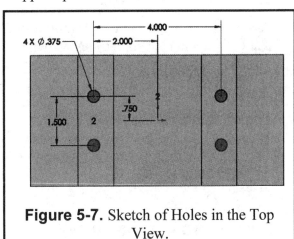

Figure 5-7. Sketch of Holes in the Top View.

Select the **Top** surface of the **Base** and click on the **Top** view orientation. Enter the **Sketch** mode and **Draw** a Circle **1.50 inch** in Diameter. Next, **Extrude** this circle **.25 inches upward**. Add the **0.125 inch Fillet** at the intersection between the cylinder and the base of the frame. Add a Chamfer to the two top inner edges of the frame and Fillets to the inner edges of the base. These are all .25". See **Figure 5-8** for a view of the completed Frame.
In the Feature Manager Tree, right click on **Edit Material** and assign **AISI 4340 Steel, annealed** to the Frame.

You can now save the completed part. Pull down **File**, select **Save As,** and save the part as **FRAME.sldprt** in your folder. Now **Close** your Frame part. You are now ready to start the next part for the assembly.

WING BASE

Go to your folder and open **ANSI-INCHES.prtdot**. Immediately pull down **File** again, select **Save As**, and then name it **WING BASE.sldprt**. Select the **Front** plane icon in the Feature Manager tree and select the **Front** view orientation. Go to **Tools – Options – Document Properties** to set **Grid/Snap** major grids to **1** and minor grids per major to **8**. Be sure to check (√) "Display Grid" **On**. **The grid will not show until you begin the sketch.**

Review the beginning sketch details shown on **Figure 5-9**. Enter the **Sketch** mode and use the **Line** tool to draw the outline. Add four sketch **Fillets** to the four corners with radii (**0.125** or **0.500**) as indicated. Draw a vertical centerline through the origin. **Select** all the geometry simultaneously (they all highlight cyan) and **Mirror** the half profile about the centerline.

Click the **Extrude Boss/Base** feature icon and use a **Mid Plane** end condition with **2.500** inches distance. Then click the green (√) button to execute the extrusion for the base part, as shown in **Figure 5-11**. Rename this "Base-Extrude" feature as **Base Part** in the Feature Manager.

Click on the top surface of the "Base Part." It should turn *blue*. Also choose a **Top** view orientation. **Sketch a Circle** on this surface, centered at the origin, with a diameter of **1.250"**. Click the **Extrude Boss/Base** feature icon. Set "Direction 1" (up) to **Blind**

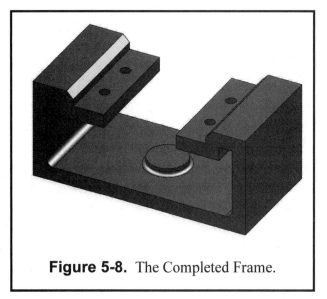

Figure 5-8. The Completed Frame.

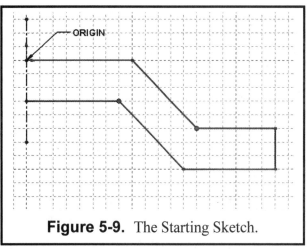

Figure 5-9. The Starting Sketch.

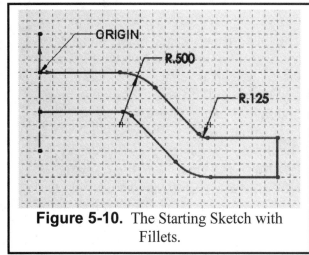

Figure 5-10. The Starting Sketch with Fillets.

with a distance of **0.25** inch. Set "Direction 2" (down) to **Blind** with a distance of **1.00** inch. Then click the green (√) button. Rename this feature **Big Boss**. Click on the top surface of this big boss, **Sketch** a **Circle** on this surface, centered at the origin, with a diameter of **0.75**. Click the **Extrude Cut** feature icon and cut the hole using the **Through All** end condition. Rename this new hole feature **Big Hole**. You now have a model that looks like **Figure 5-11** in a **Rotated View** orientation.

You now need to add a 3D chamfer and fillets to some of the edges just created. Study **Figure 5-12** to determine which edges to chamfer and fillet. To better view these edges, change to a **Wireframe** view. Click the **Chamfer** icon on the features toolbar, pick the edge to chamfer with the cursor (it turns *cyan*), and set the following parameters on the "Chamfer" menu:

> Dot (•) "Angle Distance"
> Distance = **0.10** inches
> Angle = **45** degrees

Then click the green (√) button.

Next, click the **Fillet** icon on the features toolbar, and pick the two edges where the cylinder intersects the base to fillet (they turn *blue*). Set the fillet radius to **0.125** inches. Then click the green (√) button. You can switch back to a **Shaded, Trimetric** view to better see the chamfer and fillets as shown in **Figure 5-13**.

The final features to add to the Wing Base are the four holes on the two wings where the rivets will be attached. Click on the top surface of the left wing feature to highlight it *blue*. Also select a **Top** view orientation. Start a **Sketch** and draw a **Circle** on the surface as indicated by **Figure 5-14**. Go to **Tools – Options – Document Properties** and unselect the Grid Display. Use the **Dimension** tools to precisely locate this first hole.

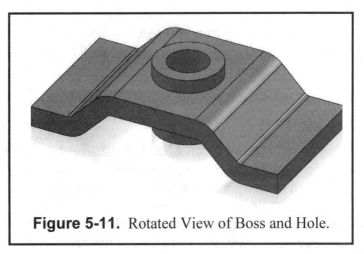

Figure 5-11. Rotated View of Boss and Hole.

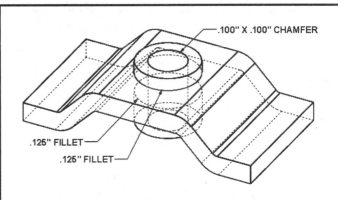

Figure 5-12. The Edges to Chamfer and Fillet.

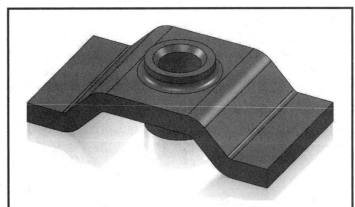

Figure 5-13. Added the Chamfer and Fillets.

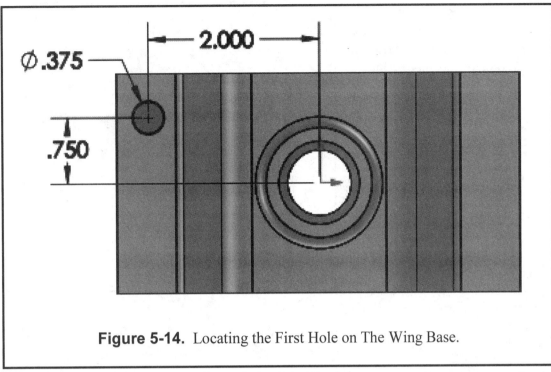

Figure 5-14. Locating the First Hole on The Wing Base.

Now select **Tools**, **Sketch Tools**, and **Linear Sketch Pattern**. In the menu, supply the following parameters:

Direction 1

Number = **2**   Spacing = **4.000** inches   Angle = go to the **right**

Direction 2

Number = **2**   Spacing = **1. 500** inches   Angle = go **down**

Click **Preview** (see **Figure 5-15**) to see if they are correct and then click **OK** to execute the command.

Click the **Extrude Cut** feature icon and cut the holes using the **Through All** end condition. Rename this new feature **Four Holes**. You now have a finished model of the Wing Base that looks like **Figure 5-16** in a **Trimetric** view orientation.

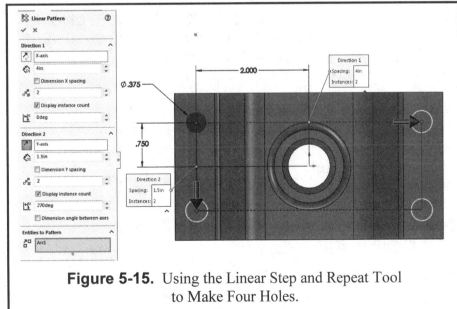

Figure 5-15. Using the Linear Step and Repeat Tool to Make Four Holes.

The Wing Base is made of Chrome Stainless Steel material properties that can be reflected by the color of the shaded image. In the Feature Manager Tree, right click on **Edit Material** and assign **Chrome Stainless Steel** to the Wing Base.

You can now save the completed Wing Base. Pull down **File**, select **Save As**, and save the part as **WING BASE.sldprt** in your folder. Now **Close** your Wing Base part. You are now ready to start the next part for the assembly.

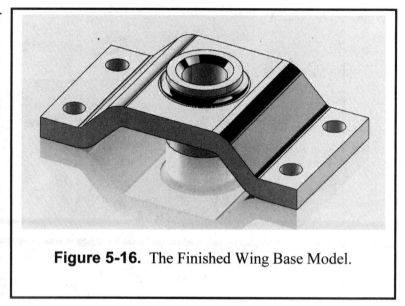

Figure 5-16. The Finished Wing Base Model.

PIN

Go to your folder and open **ANSI-INCHES.prtdot**. Immediately pull down **File** again, select **Save As**, and then name it **PIN.sldprt**. Next, click the **LMB** on the **Front** plane icon in the Feature Manager tree and also click the **Front** view orientation. Set **Grid/Snap** major grids to **1** and minor grids per major to **8.** Be sure to check (√) "Display Grid" **On**.

Review the beginning sketch details shown on **Figure 5-17**. Enter the **Sketch** mode and use the **Line** tool to draw the outline. Draw a vertical **Centerline** through the **ORIGIN**. Add the sketch **Fillet** to the corner with a radius of **0.125** as indicated. Click the **Revolve Boss/Base** feature icon and revolve the sketch profile **360** degrees. Rename this base part as **Pin Base** in the Feature Manager.

You now need to add the hex head feature to the top of the Pin. Click on the top flat surface of the Pin (it turns *blue*) and also select a **Top** view orientation. Pull down **Tools, Sketch Entities,** and then select **Polygon**. Center the polygon at the origin and drag a corner point horizontally out to the right and click the corner on the perimeter edge of the Pin, as shown in **Figure 5-18**.

Select an **Isometric** view orientation to better see the next operation. Click the **Extruded Cut** feature icon, and set the end condition to **Through All**. Also check (√) **on** the **Flip side to cut** so that the

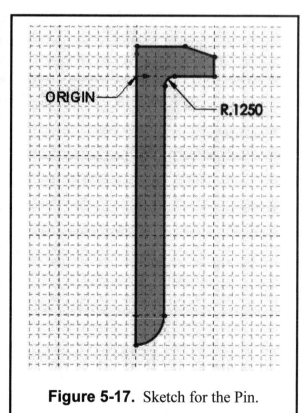

Figure 5-17. Sketch for the Pin.

six slivers of the hex head will be cut away (see **Figure 5-19**). Click the green check (√) to close the "Cut-Extrude" menu and the hex head is cut along the rim of the Pin. Rename this feature to **Hex Cut** in the Feature Manager.

In the Feature Manager Tree, right click on **Edit Material** and expand **SOLIDWORKS DIN Materials**. Then expand **DIN Copper Alloys** and assign **2.1020 (CuSn6)** to the Pin. The Pin is now complete as can be seen in **Figure 20**.

You can now save the final part. Pull down **File**, select **Save As**, and save the part as **PIN.sldprt** in your folder. Now close your session, pull down **File** and click **Close**. You are now ready to start the next part for the assembly.

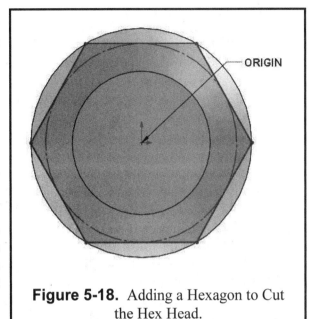

Figure 5-18. Adding a Hexagon to Cut the Hex Head.

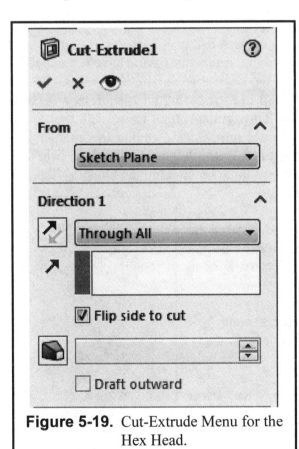

Figure 5-19. Cut-Extrude Menu for the Hex Head.

Figure 5-20. The Finished Pin Model.

RIVET

The Rivet is a simple part and you can create it in a number of ways. The method chosen here is to simply sketch a half-profile and then revolve it around a centerline. There are different types of rivets (flat head, pan head, etc.). For this case, you will design a "Button Head" rivet.

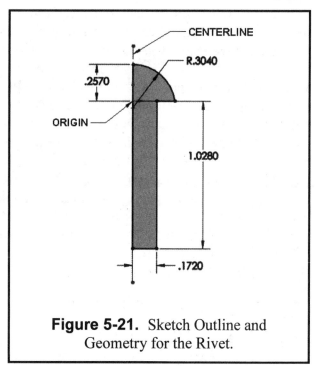

Study the geometry in **Figure 5-21**. Go to your folder and open **ANSI-INCHES.prtdot**. A good practice is to immediately name your part. Pull down **File** again, select **Save As**, and then name it **RIVET.sldprt**. Select the **Front** plane icon in the Feature Manager tree and also click the **Front** view orientation. Disregard the "Grid/Snap" settings, since you will just sketch an outline and then fix the geometry with the dimensioning tool.

Figure 5-21. Sketch Outline and Geometry for the Rivet.

Enter the **Sketch** mode and use the **Line** and **Centerpoint Arc** tools to draw the outline as indicated in **Figure 5-21**. Also draw a vertical **Centerline** through the **ORIGIN**. Then use the **Dimension** tool to set all the given dimensions. If they are placed correctly, all lines should turn *black* meaning the geometry is fixed. **Note:** **The center of the arc is not at the origin.** If it happens to be at the origin in your case, then remove the "Coincident" relation between them by clicking the **Display/Delete Relations** icon (it looks like a pair of glasses).

Now click the **Revolve Boss/Base** feature icon and revolve the sketch profile **360** degrees. Rename this base part as **Rivet Base** in the Feature Manager. The finished rivet is shown in **Figure 5-22** in an Isometric View.

In the Feature Manager Tree, right click on **Edit Material** and assign **2.0090 (Cu-DHP)** from the DIN Copper Alloys Materials list to the Rivet.

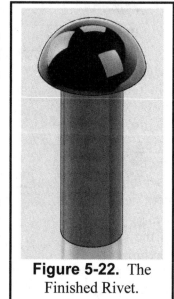

You can now save the completed Rivet. Pull down **File**, select **Save As**, and save the part as **RIVET.sldprt** in your folder. Now close your session, pull down **File** and click **Close**. You are now ready to start the assembly.

Figure 5-22. The Finished Rivet.

TERMINAL SUPPORT ASSEMBLY

Pull down **File**, **New** and select **Assembly**. Pull down **File** again, select **Save As**, and then save it as **TERMINAL SUPPORT**. Pull down **File** and **Open,** one-by-one, the four parts of the assembly: **Frame, Wing Base, Pin**, and **Rivet**. You now have five active files in your current SOLIDWORKS session. Pull down **Window** and select **Tile Vertically**. You can now see all four files on the screen as shown in **Figure 5-23**.

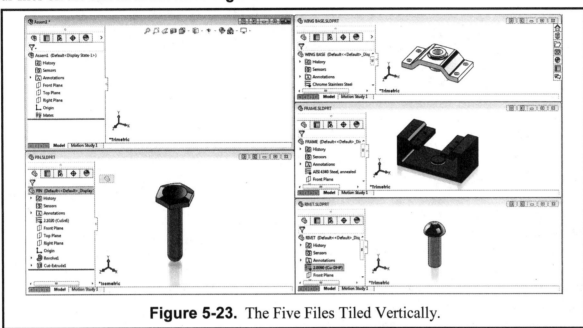

Figure 5-23. The Five Files Tiled Vertically.

Starting with the **Frame** file, **Select** its title in the Feature Manager and "Drag and Drop" it into the assembly file. **NOTE**: The first part dropped into the assembly is **FIXED** and becomes the base to which all other parts are mated. It cannot be moved unless its constraints are deleted. Repeat this "Drag and Drop" with the **Wing Base** then the **PIN**. Finally, "Drag and Drop" the **RIVET** <u>**four times**</u> into the assembly file. You can try to line them up when you "Drop" them into the assembly, but more importantly you will mate them next. You no longer need the four part files so you can **Close** them by clicking the X on the top right side of their window. Then maximize the assembly file window. You should now have a computer screen layout similar to **Figure 5-24**.

Figure 5-24. The Parts after Dropping them into the Assembly.

You will mate the Frame and the Wing Base first. Click the **Assembly Tab** and select the **Mate** icon (it looks like a paper clip), and the "Mate" menu comes up as indicated in **Figure 5-25**. Now pick the cylindrical hole on the Frame and the corresponding hole in the Wing base. Do not select the edges of the holes. Click on the **Concentric** mate setting. Repeat this process for one more set of corresponding holes. This will constrain the Wing Base with the Frame along the axes of the holes. The next operation will establish constraints in relation to the top surface of the Frame and the bottom surface of the upper wing of the Wing Base. Select the **Mate** icon and select the top surface of the Frame. Rotate the figure up so you can see the bottom surface of the Wing Base and select the corresponding surface. Click on the **Coincident** mate setting.

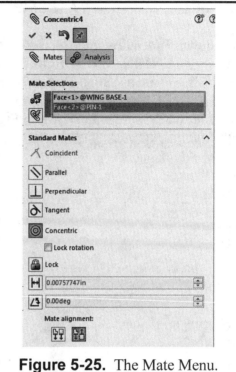

Figure 5-25. The Mate Menu.

Now pick the outer cylindrical surface of the Pin (the surface that will go into the hole) and then pick inner cylindrical surface of the big hole on the Wing Base. They both should turn *blue*. Now click the **Concentric** mate setting and then click the green check (√) to execute the mate. The Pin is now concentric with the hole and you should see it move over a little. You can look at it from a **Top** view if you wish to check it.

Repeat this **Concentric** mate four times, one-by-one, between each Rivet and its respective small hole on the wing surfaces. Use an **Isometric** view and **Zoom** in as needed to complete these mates.

Now try something. Click the **Move Component** icon on the left side Assembly toolbar. Try to move the Pin or one of the Rivets around on the screen. You will see that its movement is constrained in a vertical direction that corresponds to the axis of its respective hole.

Figure 5-26. Using a Rotated View to See the Bottom Surface of the Hex Head.

You will now fix the parts to the Wing Base. Click the **Mate** icon again. Click the flat surface on the lip of the big hole. Next, you need to click the bottom flat surface of the hex head, but it may be hard to select. So select the **Rotate View** icon and rotate the whole assembly around until you have a clear view of that surface. Now de-select the "Rotate View" icon (by clicking it off) to get back the pick cursor, and select the bottom of the hex head as shown in **Figure 5-26**.

Now on the "Mate" menu, click the **Coincident** mate setting and then click the green check (√) to execute this type of mate. You should see the Pin move into the correct position on the Wing Base and it is now fixed. Return to an **Isometric** view and use the **Move Component** icon to see if you can move the Pin up and down inside the big hole. It is fixed and you should not be able to move it out of the hole. Now click the **Rotate Component** icon on the Assembly toolbar. You should be able to rotate it around in the hole since that degree-of-freedom is not fixed. Now repeat this **Coincident** mating process for the four rivets, one-by-one. In each case, mate the flat bottom surface of the Rivet head with the top flat wing surface of its respective hole. Use the **Rotate View** feature as needed (or if you think you can get away with it, use a **Wireframe** view mode). After you have completed these four mates, try the **Move Component** and **Rotate Component** tools until you are satisfied that the assembly has been mated correctly. You are now finished with this mating process and your Terminal Support Assembly is complete.

Now pull down **File**, select **Save As**, and save the assembly as **TERMINAL SUPPORT.sldasm** in your folder. Insert the rendered Terminal Support Assembly image into your **Title Block** drawing sheet that was created in Chapter 1. You will now create a Bill of Materials and add Balloons with Part names attached to the assembly. To begin, go to Tools – Options – Document Properties and make the changes that are shown in **Figure 5-27**. Set Leader style and Frame style to **0.0098in**. The leader display for Single/Stacked Balloons and Auto Balloons should both be Bent. Under the Single balloon selection box, set the style to Circular and the size to 2 Characters. Now click on **Font**. And set the Font to **Arial – Regular** – and the **Height to 0.25in**. Once all of these settings are made, click the **OK** tab.

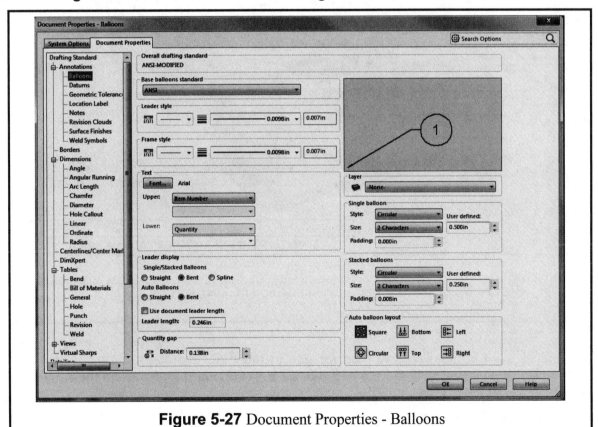

Figure 5-27 Document Properties - Balloons

Select **Insert–Annotations – Balloon** and carefully select the Frame, the Wing Base, the Pin and one Rivet. Circles with the respective Part numbers will appear. After all of the Balloons are placed, select **Insert Annotations Note** (without a Leader) and place the part name (18 pt) next to the corresponding balloon.

Now select **Insert – Tables – Bill of Materials.** Once you select the assembly, a selection menu will appear where the Feature Manager tree is. Select Parts only, display all configurations of the same part as one item; for the border, the first number should be changed to 0.0098 and the second number is unchanged. Okay your selections and a Bill of Materials will appear on the drawing sheet. Place it in the upper left corner of your drawing

Now save your drawing as **TERMINAL SUPPORT.slddrw** and **Print** it on this sheet (see **Figure 5-28**). Now close your session, pull down File and click close. You are finished.

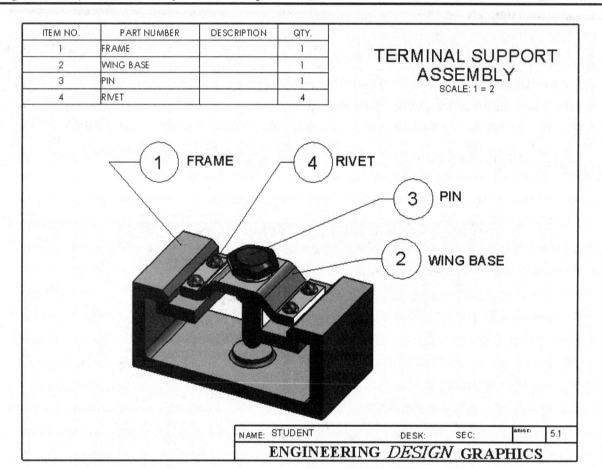

ITEM NO.	PART NUMBER	DESCRIPTION	QTY.
1	FRAME		1
2	WING BASE		1
3	PIN		1
4	RIVET		4

TERMINAL SUPPORT ASSEMBLY
SCALE: 1 = 2

NAME: STUDENT   DESK: SEC: 5.1

ENGINEERING *DESIGN* GRAPHICS

Figure 5-28. The Finished Terminal Support Assembly on the Title Sheet.

Be sure to save your assembly files for use in later lab exercises such as "Kinematics Simulation" where you will need to explode an assembly and make an animation file.

Exercise 5.2: PULLEY ASSEMBLY

The Pulley Assembly is a common design used to hoist objects with a mechanical leverage advantage. It has four major part components: an Eye Hook, a Pulley Sheave, Spacer, and a Base Plate that is used twice. It is assembled using small rivets that are peened on one end to secure them and hold the components together. You will start the design by designing the Base Plate, which controls many of the dimensions of the other components.

BASE PLATE

Study the geometry of the Base Plate in **Figure 5-29**. Go to your folder and open **ANSI-INCHES.prtdot**. Immediately pull down **File** again, select **Save As**, and then name it **BASE PLATE.sldprt**. Next, click the **LMB** on the **Front** plane icon in the Feature Manager tree and also click the **Front** view orientation. Your sketch will be symmetrical about a vertical centerline. Disregard the "Grid/Snap" settings, since you will just sketch an outline and then use the dimensioning tool.

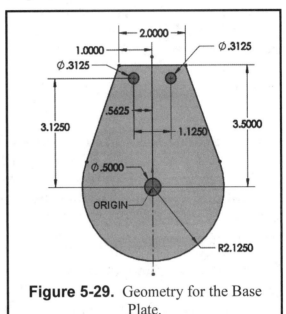

Figure 5-29. Geometry for the Base Plate.

Enter the **Sketch** mode and use the **Line** and **Circle** sketch tools to draw the outline as indicated in **Figure 5-29**. Start with a **Circle** at the origin and with a radius of **2.125** inches. Then draw three **Lines**, approximating the tangent points. Click the **Add Relation** icon and make the **Tangent** relation with the bottom circle and left line, then make the **Tangent** relation with the bottom circle and right line. Use **Trim** to get rid of part of the circle not needed. Draw the two small **Circles** on the top of the profile, draw the center **Circle**, and make sure the top line is **Horizontal**. Finally, use the **Dimension** tool to fix all the geometry in place. If done correctly, the lines should turn *black*. Select **Extrude Boss/Base**, use a **Blind** end condition and distance of **0.1250** inches (one-eighth inch). In the Feature Manager Tree, right click on **Edit Material** and assign **AISI 4340 STEEL,**

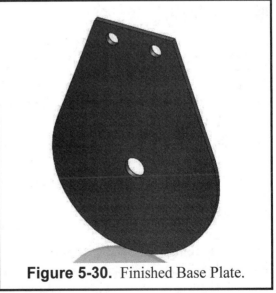

Figure 5-30. Finished Base Plate.

NORMALIZED from the SOLIDWORKS Materials – Steel list. Then click the **Apply and Close buttons**. The Base Plate is finished as shown in **Figure 5-30**.

Be sure to **Save** your **BASE PLATE.sldprt** before you **Close** your file and then start a new part for the assembly.

PULLEY

Study the geometry of the Pulley in **Figure 5-31**. Go to your folder and open **ANSI-INCHES.prtdot**. Immediately pull down **File** again, select **Save As**, and then name it **PULLEY.sldprt**. Next, click the **LMB** on the **Right Plane** icon in the Feature Manager tree and also click the **Right** view orientation. Disregard the "Grid/Snap" settings for this part.

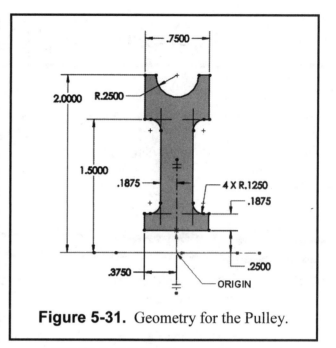

Figure 5-31. Geometry for the Pulley.

Enter the **Sketch** mode and use the **Line** and **Centerpoint Arc** sketch tools to draw the outline as indicated in **Figure 5-31**. Also draw a **Vertical and Horizontal Centerline** through the **Origin**. The outline is symmetrical about the vertical centerline so you can use the Dynamic Mirror function found under **Tools – Sketch Tools**. Use the **Dimension** tool to fix all the geometry in place. If done correctly, the lines should turn *black*. Now select the **Revolve Boss/Base** feature icon and give the revolution a full **360** degrees parameter. You now have the pulley part of the assembly.

In the Feature Manager Tree, right click on **Edit Material** and expand **SOLIDWORKS DIN Materials**. Then expand **DIN Copper Alloys** and assign **2.0375 (CuZn36Pb3)**. Then click the **Apply and Close buttons**. The Pulley is finished as shown in **Figure 5-32**. Be sure to **Save** your **PULLEY.sldprt** before you **Close** your file.

Figure 5-32. The Finished Pulley.

SPACER

Study the geometry of the Spacer in **Figure 5-33**. Go to your folder and open **ANSI-INCHES.prtdot**. Immediately pull down **File** again, select **Save As**, and then name it **SPACER.sldprt**. Next, click the **LMB** on the **Front Plane** icon in the Feature Manager tree and also click the **Front** view orientation. Disregard the "Grid/Snap" settings for this part.

Enter the **Sketch** mode and use the **Line** and **Circle** sketch tools to draw the simple outline as indicated in **Figure 5-33**. Then use the **Dimension** tool to fix all the geometry in place. Be sure to dimension the position of the Spacer relative to the <u>origin</u>. If done correctly, the lines should turn *black*.

Select **Extrude Boss/Base,** use a **Mid Plane** end condition and distance of **1.000** inches. *Note*: A Mid Plane end condition will extrude the sketch **0.500** inches in both directions, using the sketch plane as the "mid plane."

Now click on the top surface of the Spacer and **Sketch** a **Circle** centered at the origin and with a <u>diameter</u> of **0.3500** inches, as shown in **Figure 5-34**. Select **Extrude Cut** and cut the hole **Through All** the Spacer.

In the Feature Manager Tree, right click on **Edit Material** and assign **AISI Type A2 Tool Steel** from the SOLIDWORKS Materials – Steel list. Then click the **Apply and Close buttons**. The Spacer is finished. Be sure to **Save** your **SPACER.sldprt** before you **Close** your file.

EYE HOOK

The Eye Hook part can be made by simply revolving a circle and then extruding a circle in their respective planes. Go to your folder and open **ANSI-INCHES.prtdot**. Immediately name your part by pulling down **File** again, select **Save As**, and then name it **EYE HOOK.sldprt**. Next, click the **LMB** on the **Right** plane icon in the Feature Manager tree and also click the **Right** view orientation. Set the **Units** to **Inches** and **4** decimal places. Disregard the "Grid/Snap" settings for this eye hook part.

Sketch a **Circle** in the position indicated in **Figure 5-35**. Also draw a horizontal **Centerline** above this circle. Click the **Add Relations** icon, and make the "Circle" and "Origin" have a **Vertical** relation. Add a **Horizontal** relation to the "Centerline" just to be safe. Now use the **Dimension** tool to add the three dimensions that will fix all the geometry.

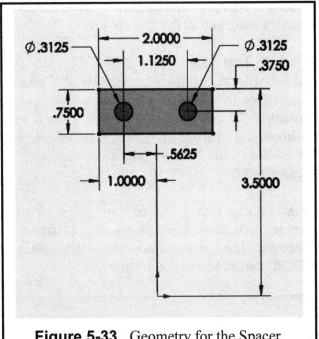

Figure 5-33. Geometry for the Spacer.

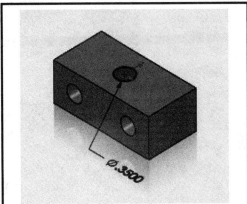

Figure 5-34. Adding the Eye Hook Hole to the Spacer.

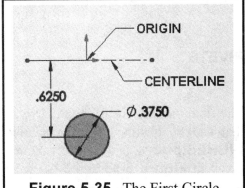

Figure 5-35. The First Circle Position for the Eye Hook.

Select the **Revolve Boss/Base** icon and revolve the circle through a **360** degrees path. This creates the top part of the Eye Hook.

Select the **Top** plane in the Feature Manager. Pull down **Insert**, select **Reference Geometry**, and then **Plane**. Define the plane to be **.625** inches **Below** the **Top plane**, and then click the green check (√) mark. Now click on this new plane (called **Plane 1** in the Feature Manager). **Sketch** a **Circle** on it centered at the plane's <u>local origin</u> and with a <u>diameter</u> of **0.3475** inches, as shown in **Figure 5-36**. Next, **Extrude Boss/Base** the circle to a **Blind** distance of **1.2500** inches downward. This completes the geometric construction of the Eye Hook as shown in **Figure 5-37**.

In the Feature Manager Tree, right click on **Edit Material** and assign **Tin Bearing Bronze** from the SOLIDWORKS Materials – **Copper Alloys** list. Then click the **Apply and Close buttons**. The Eye Hook now is finished and has a very nice color. Be sure to **Save** your **EYE HOOK.sldprt** before you **Close** your file.

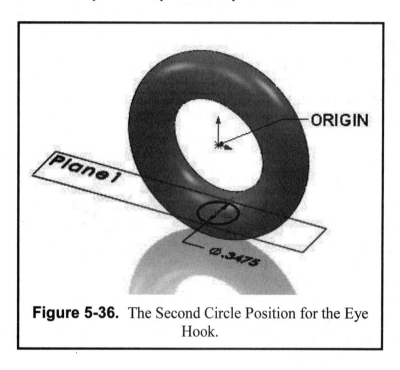

Figure 5-36. The Second Circle Position for the Eye Hook.

Figure 5-37. The Finished Eye Hook.

RIVETS

The final parts to create are the Big and Small Rivets. These are standard parts with dimensions that can be obtained from a handbook. In <u>each case</u>, go to your folder and open **ANSI-INCHES.prtdot.** and **Sketch** the given outline in a **Front** plane. Use the **Dimension** values given in **Figure 5-38** and **5-39** for the Big Rivet and Small Rivet, respectively. Center your data according to the given origin. <u>Note:</u> **Do not draw the arcs for the rivets at the origin.** When you **Dimension** the **Centerpoint Arc** entity, the center of that arc moves down slightly below the origin. Be sure to draw a vertical **Centerline** through the origin. **Revolve Boss/Base** the Rivet outline **360** degrees. In the Feature Manager Tree, right click on **Edit Material** and assign **2.0060 (Cu-ETP)** from the SOLIDWORKS DIN Materials – DIN Copper Alloys list. Then click the **Apply and Close buttons**. **Save** the Big Rivet as **BIG RIVET.sldprt** before you **Close** your file. **Save** the Small Rivet as **SMALL RIVET.sldprt** before you **Close** your file.

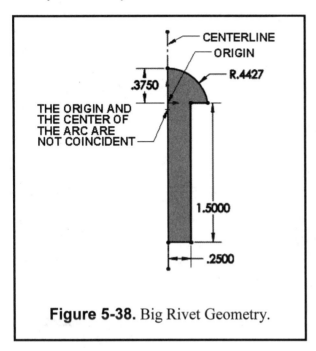

Figure 5-38. Big Rivet Geometry.

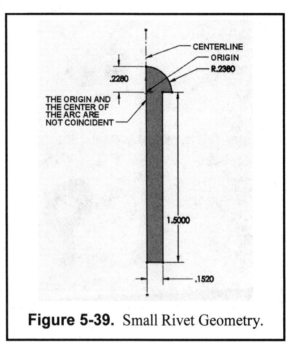

Figure 5-39. Small Rivet Geometry.

SWIVEL EYE BLOCK ASSEMBLY

Pull down **File**, **New** and select **Assembly**. Pull down **Tools, Options, Document Properties** and select **ANSI** and then Select **Units**. Make sure you indicate **IPS** (Inch, Pound, Second) and **OK** the selections. Pull down **File** again, select **Save As**, and then save it as **SWIVEL EYE BLOCK.sldasm**. Pull down **File** and **Open,** one-by-one, the four main parts of the assembly:

> **BASE PLATE.sldprt**
> **PULLEY.sldprt**
> **SPACER.sldprt**
> **EYE HOOK.sldprt**

You will add the Rivets later. You now have five active files in your current SOLIDWORKS session (the Assembly file and the four Part files). Pull down **Window** and select **Tile Vertically.** You can now see all five files on the screen as shown in **Figure 5-40**.

NOTE: The first part dropped into the assembly is **FIXED** and becomes the base to which all other parts are mated. It cannot be moved unless its constraints are deleted. Starting with the **SPACER** file, select its title in its Feature Manager and "Drag and Drop" it into the assembly file. Repeat this "Drag and Drop" with the **PULLEY** and then the **EYE HOOK**. Finally, "Drag and Drop" the **BASE PLATE** <u>two times</u> into the assembly file. You can try to line them up when you "Drop" them into the assembly, but more importantly you will mate them next. You no longer need the four part files so you can **Close** them by clicking the X on the top right side of their window. Then maximize the assembly file window. You should now have an **Isometric** computer screen layout like **Figure 5-41**.

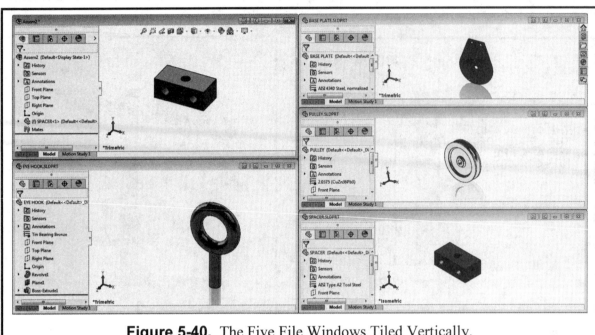

Figure 5-40. The Five File Windows Tiled Vertically.

You will first mate the Front Base Plate with the Spacer. Click the **Assembly Tab** and select the **Mate** icon (it looks like a paper clip), and the "**Mate**" menu comes up as indicated in **Figure 5-42**. Now pick the upper left small cylindrical hole surface of the front Base Plate and then pick the inner left cylindrical hole surface of the Spacer. They both should turn *blue*. Now click the **Concentric** mate setting and then click the green check (√) to execute the mate. Repeat this **Concentric** mate process by picking the upper right small cylindrical hole surface of the Base Plate and then pick the inner right cylindrical hole surface of the Spacer. Use an **Isometric** view and **Zoom** in as needed to perform these mates.

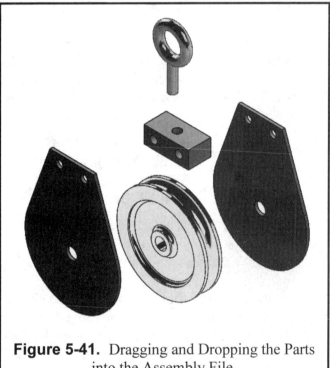

Figure 5-41. Dragging and Dropping the Parts into the Assembly File.

Now their holes are concentric but the front Base Plate and Spacer may not touch. Click the **Mate** icon and then pick the front surface of the Spacer and the back surface of the front Base Plate. (*Note:* You probably cannot find a view orientation that allows you to see and pick both of these surfaces simultaneously, so use the **Rotate View** icon or the **Wireframe** view mode to assist you in clicking these two surfaces.) Once you have selected the two flat surfaces, they should turn *blue*. Now in the "Mate" menu, select the **Coincident** mating condition, and click the green (√) **OK** button. You should now see the front Base Plate move towards the front surface of the Spacer, and they are now mated properly.

Repeat this process with the back side of the Spacer and the back Base Plate. This includes three **Mates**:

- o **Concentric** holes on left side
- o **Concentric** holes on right side
- o **Coincident** surfaces that touch

The two Base Plates are now mated to the Spacer as shown in an **Isometric** view in **Figure 5-43**. You can look at it from a **Front** view orientation if you wish to check it and see the holes going all the way through as shown in **Figure 5-44**.

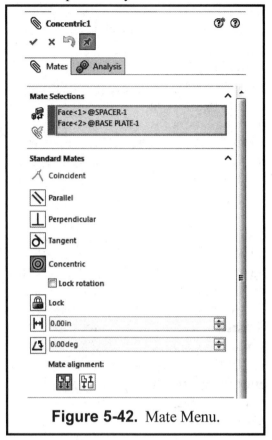

Figure 5-42. Mate Menu.

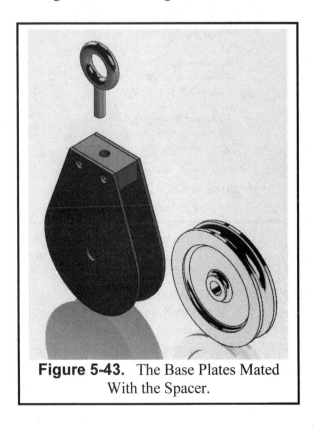

Figure 5-43. The Base Plates Mated With the Spacer.

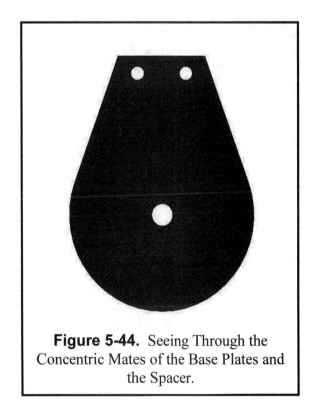

Figure 5-44. Seeing Through the Concentric Mates of the Base Plates and the Spacer.

Now you will mate the Pulley into the Base Plates that will sandwich it. Click the **Mate** icon. Now pick the large center hole surface of the front Base Plate (*or* you could *instead* pick the large center hole surface of the back Base Plate). Next, pick the center hole surface of the Pulley. Give it a **Concentric** mate and then click the green (√) button. Notice the Pulley aligns concentrically with the center Base Plate holes but in all likelihood is not sandwiched exactly between them. Switch to a **Right** view orientation to better see this. You could just move it into position by clicking the **Move Component** icon and then moving the Pulley into position by "eyeball." However, that is not the right way to do it. You need a second mating condition.

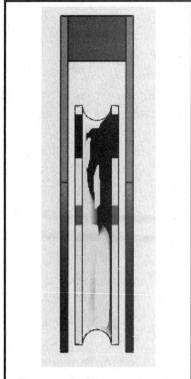

The Spacer is 1.000 inches thick and the pulley is 0.750 inches thick. So you need to position the Pulley (1.000-0.75)/2 = 0.125 inches away from the Base Plates. Click the **Mate** icon. Select the front round face of the Pulley and then select the back face of the front Base Plate. You may need to use the **Zoom** and **Rotate View** icons to get these faces highlighted simultaneously. Once they are highlighted (turned *blue*), select the **Distance** mate setting, key in the value of **0.1250**, and then click the green (√) button. The Pulley moves into position and is now fully mated. Use a **Right** view orientation to see this latest mate, as shown in **Figure 5-45**. Make sure there is a small gap.

Figure 5-45. Mating the Pulley between the Two Base Plates.

While you are at it, notice the "Mates" branch in your Feature Manager tree structure. It lists all the mates made to date for the Swivel Eye Block Assembly, as shown in **Figure 5-46**. You can "right click" on the mate of interest and **Edit Feature** of the mate.

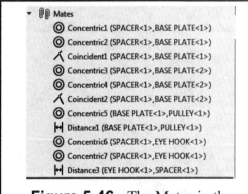

Figure 5-46. The Mates in the Feature Menu.

Now you will mate the Eye Hook to the assembly. Return to an **Isometric** view. Click the **Mate** icon. Select the circular shaft of the Eye Hook and the top circular hole on the Spacer. The two surfaces turn *blue*. Now give them a **Concentric** mate and click the green (√) button. The shaft now mates with the hole, but it probably does not go far enough down the hole. You can switch to a **Right** view and **Zoom** into the area where the shaft goes through the Spacer hole. Use the **Move Component** assembly feature and move the shaft through the hole until it sticks out of the bottom of the Spacer. Use the **Rotate View** display icon and rotate your view until you can see both the flat bottom of the shaft and the bottom surface of the Spacer. Click **Mate** and then click both the flat bottom of the shaft and the bottom of the Spacer surfaces. Give them a **Distance** mate of **0.1250** inches. **Preview** the selection (see **Figure 5-47**) and if it is correct, then click the green (√) button. If it does not look correct, you can try **Flip Dimension**.

The major parts of the assembly are now mated completely, and you will next add the Rivets. First though, take note of the degrees-of-freedom that still remain with these parts. The two Base Plates and the Spacer are fixed in space because of the two concentric and one coincident mate conditions. However, the Pulley and Eye Hook can still rotate in one degree-of-freedom. Select the **Rotate Component** icon on the assembly toolbar. Now with the cursor, move the Pulley around its concentric mate and see it turn effortlessly. Now move the Eye Hook as it swivels around in its hole, but you cannot move it out of the hole because of the distance mate. Try to leave the Eye Hook in a position such that the Eye feature is parallel to the Base Plate, as shown in the final **Figure 5-47**.

Pull down **File** and **Open** both the **BIG RIVET** and **SMALL RIVET** files. Pull down **Windows** and select **Tile Vertically**. Drag and drop the "Big Rivet" into the assembly window, and then drag and drop the "Small Rivet" into the assembly window <u>twice</u>. You may want to use the **Move Component**, **Zoom** and **Rotate View** icons to get these faces highlighted simultaneously when you are mating (see **Figure 5-48**). Each rivet will fit into its respective hole and will have the same two identical **Mate** characteristics. The first characteristic will be a **Concentric** mate between the rivet shaft and the inner surface of its respective hole. The second mate is a **Coincident** mate between the back flat part of the rivet head and the front face of the front Base Plate. So now execute these mates for the Big Rivet going through the center holes and then for the two Small Rivets going through their respective holes on the top of the assembly. Once you are finished, you should have a final Swivel Eye Block Assembly as shown in **Figure 5-49** in a **Trimetric** view.

<u>*Manufacturing Note*</u>: If you **Rotate View** and look at the back of your assembly, you can see the Rivets sticking out about 0. 25 inches (one fourth of an inch). The normal process used to secure the Rivet to the Base Plate is a peen operation. *Peening* the end of the Rivet with a hammer, for example, causes the surface to expand in a spherical direction, and thus prevents the Rivet from coming out of the hole. We will leave this peening operation for a later day, and your assembly modeling and mating exercise is over.

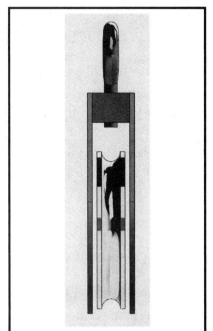

Figure 5-47. Mating the Eye Hook with the Spacer.

Figure 5-48. Dragging and Dropping the Rivets into the Assembly and rotating them.

Pull down **File**, select **Save As**, and save the assembly as **SWIVEL EYE BLOCK.sldasm** in your folder.

Before you close this file you should save it and also get a hard copy printout. Insert the rendered Swivel Eye Block Assembly image into your **Title Block** drawing sheet that was created in Chapter 1.

You will now create a Bill of Materials and add Balloons with Part Names attached to the assembly. To begin, go to Tools – Options – Document Properties and make the changes that are shown in **Figure 5-27**. Set Leader style and Frame style to **0.0098in**. The leader display for Single/Stacked Balloons and Auto Balloons should both be bent. Under the Single balloon

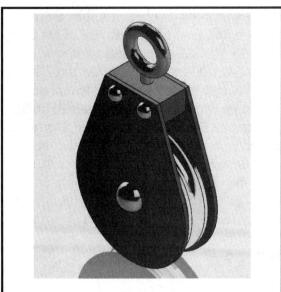

Figure 5-49. The Finished Swivel Eye Block Assembly in an Isometric View.

selection box, set the style to Circular and the size to 2 Characters. Now click on **Font**. And set the Font to **Arial – Regular** – and the **Height to 0.25in**. Once all of these settings are made, click the **OK** tab.

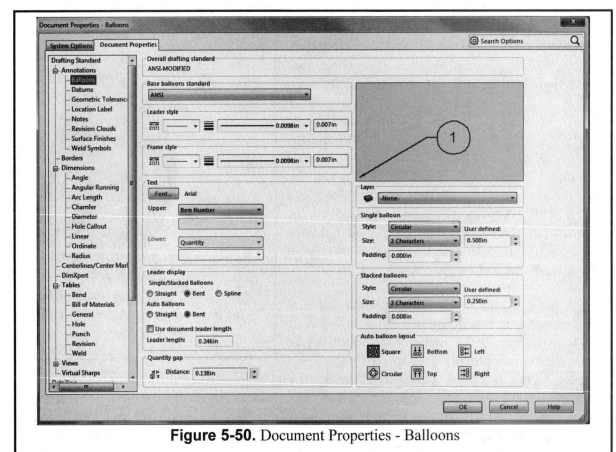

Figure 5-50. Document Properties - Balloons

Select **Insert–Annotations – Balloon** and carefully select the Spacer, one Base Plate, the Pulley and Eye Hook, one Small Rivet and the Big Rivet. Circles with the respective Part numbers will appear. After all of the Balloons are placed, select **Insert Annotations Note** (without a Leader) and place the part name (18 pt) next to the corresponding balloon.

Now select **Insert – Tables – Bill of Materials.** Once you select the assembly, a selection menu will appear where the Feature Manager tree is. Select Parts only, Display all configurations of the same part as one item; for the border, the first number should be changed to 0.0098 and the second number is unchanged. Okay your selections and a Bill of Materials will appear on the drawing sheet. Place it in the upper right corner of your drawing.

Now save your drawing as **PULLEY ASSEMBLY.slddrw** and **Print** it on this sheet (see **Figure 5-51**). If you have access to a *Color Printer*, use it to show the nice colors of your assembly.

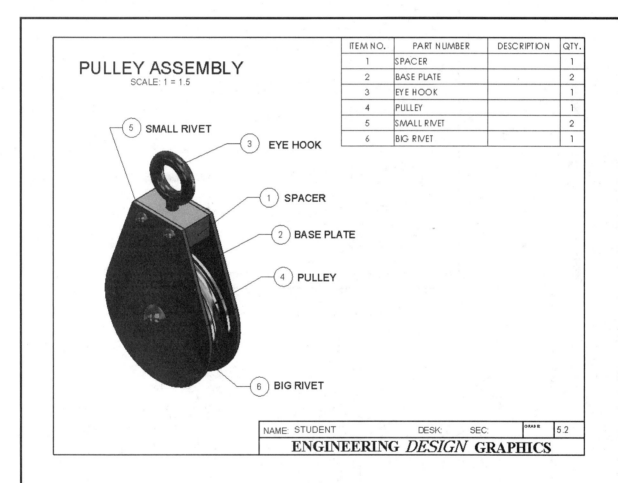

ITEM NO.	PART NUMBER	DESCRIPTION	QTY.
1	SPACER		1
2	BASE PLATE		2
3	EYE HOOK		1
4	PULLEY		1
5	SMALL RIVET		2
6	BIG RIVET		1

PULLEY ASSEMBLY
SCALE: 1 = 1.5

5 SMALL RIVET
3 EYE HOOK
1 SPACER
2 BASE PLATE
4 PULLEY
6 BIG RIVET

NAME: STUDENT   DESK: SEC: GRADE   5.2
ENGINEERING *DESIGN* GRAPHICS

Figure 5-51. The Swivel Eye Block Assembly Image Inserted Onto the Title Sheet.

Be sure to save your assembly files for use in later lab exercises such as "Kinematics Simulation" where you will need to explode an assembly and make an animation file.

SUPPLEMENTARY EXERCISE 5-1: CASTER ASSEMBLY

Create the parts of the Caster Assembly of the Option specified by your instructor in the Table provided.

1. Generate an Assembly Drawing of the Caster Parts and insert the Assembly into a Title Block.

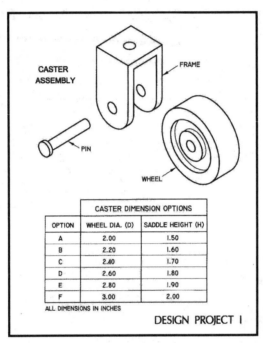

OPTION	WHEEL DIA. (D)	SADDLE HEIGHT (H)
A	2.00	1.50
B	2.20	1.60
C	2.40	1.70
D	2.60	1.80
E	2.80	1.90
F	3.00	2.00

CASTER DIMENSION OPTIONS

ALL DIMENSIONS IN INCHES

DESIGN PROJECT I

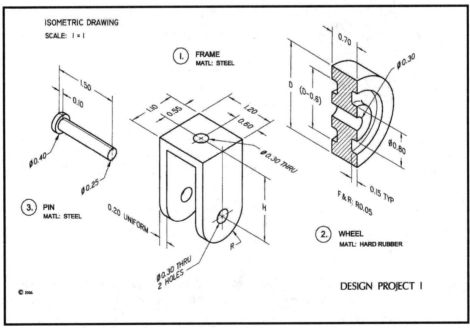

SUPPLEMENTARY EXERCISE 5-2: PULLEY ASSEMBLY

Create the parts of the Pulley Assembly of the Option specified by your instructor in the Table provided.

1. Generate an Assembly Drawing of the Pulley Parts and insert the Assembly into a Title Block.

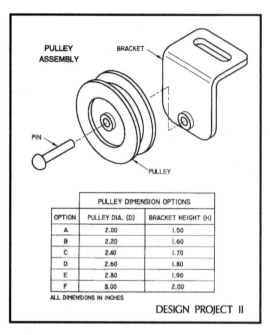

PULLEY ASSEMBLY

BRACKET

PIN

PULLEY

PULLEY DIMENSION OPTIONS		
OPTION	PULLEY DIA. (D)	BRACKET HEIGHT (H)
A	2.00	1.50
B	2.20	1.60
C	2.40	1.70
D	2.60	1.80
E	2.80	1.90
F	3.00	2.00

ALL DIMENSIONS IN INCHES

DESIGN PROJECT II

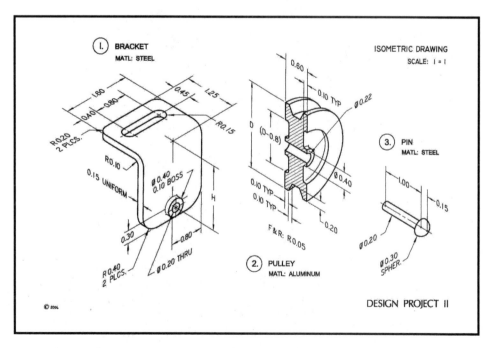

1. BRACKET
MATL: STEEL

ISOMETRIC DRAWING
SCALE: 1 = 1

0.60
0.10 TYP
Ø0.22

1.60
0.80
0.40
0.45
1.25
R0.15

R0.20 2 PLCS.

R0.10

0.15 UNIFORM

Ø0.40
0.10 BOSS

D (D-0.8)

Ø0.40

3. PIN
MATL: STEEL

1.00
0.15

H

0.10 TYP
0.10 TYP
0.20

0.30
0.80

F & R: R0.05

R0.40 2 PLCS.
Ø0.20 THRU

2. PULLEY
MATL: ALUMINUM

Ø0.20
Ø0.30 SPHER.

© 2004

DESIGN PROJECT II

NOTES:

Computer Graphics Lab 6: Analysis and Design Modification I

In Computer Graphics Lab 6, you will be introduced to some of the advanced analysis and design modification capabilities of SOLIDWORKS. Solid modeling is a tool for engineering design that offers many advantages over conventional practices. One of these advantages is the ability to analyze the design directly from the digital database without the need to build a physical part. In this laboratory, you will analyze various design properties of the solid model using the **Measure** and **Mass Properties** commands. You will also be exposed to **Design Tables**, which can be used to build a family of similar parts.

THE MEASURE FUNCTION

The **Measure** function can be found in the **Tools** pull-down menu (see **Figure 6-1**). When you use this function, a ruler appears on the cursor and you can pick the entity of interest in the computer screen area. This function includes the following capabilities as examples. They can be applied either to a 2D sketch or to a 3D solid model.

Line returns the length of the line.

Arc returns the length of the arc and the diameter. If the arc is a full circle, the length ends up being the circumference (see **Figure 6-2**).

Flat Surface returns the units square for area of the surface and the length of the perimeter around the flat surface.

Circular Surface returns the units square for area of the surface, the diameter, and the perimeter. If it is a through hole, then the perimeter will be the circumference of one end plus the circumference of the other end.

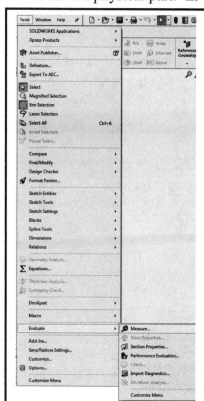

Figure 6-1. The Measure Function.

THE MASS PROPERTIES FUNCTION

Pull down **Tools, Evaluate,** and select the **Mass Properties** command. (see also **Figure 6-1**). This will open an on-screen window that reports the mass properties of the current solid model. Before you can calculate the mass properties accurately, you need to set the density of the material for the solid model part. The density is set when the material is assigned. In the Feature Manager Tree right click on "Material" and select Edit Material. Once the material is assigned the density of the material is also set, as shown in **Figure 6-3**. See also **Table 6-1** for some common density values.

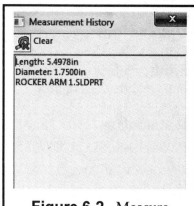

Figure 6-2. Measure Screen Menu.

Once the density of the part material is set, you can then select the **Mass Properties** command. The result will be reported in the "Mass Properties" Report in the on-screen window, as shown in **Figure 6-4**. The mass property values and units reported include the following (see **Page 6-11**, for definitions):

Part Name

Density (lbs/in³⁾

Mass (lbs)

Volume (in³)

Surface Area (in²)

Center of Mass (in)

Moments of Inertia (lbs*in²)

Once the "Mass Properties" Report appears on the screen, you can review the values reported. To see all of them at once, you can stretch the lower right corner to enlarge the on-screen window. There are also some options buttons you can click at the top of the menu.

Print will open the print menu where you can print a copy of the "Mass Properties" Report.

Copy will copy the entire report to the computer clipboard. Then you can open a word processing software like MS-Word or WordPad and paste it directly into the document. This allows you to edit the report in order to get a customized hardcopy of it.

Close will simply close the on-screen "Mass Properties" window.

Options will open another "Mass Property Options" window on the screen as shown in **Figure 6-5**. Here you can custom set the units and density values.

Recalculate will recalculate the mass properties after options are changed.

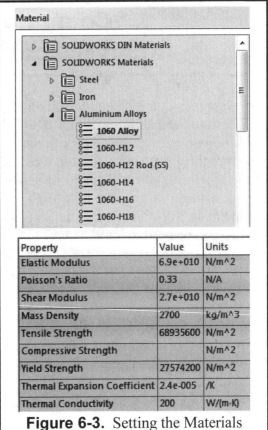

Figure 6-3. Setting the Materials Properties.

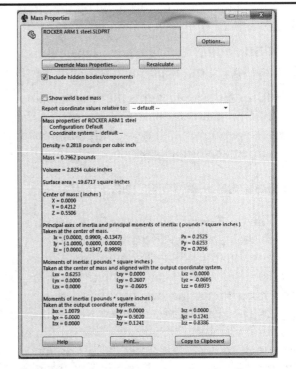

Figure 6-4. Mass Properties On-Screen Report.

ABOUT THE MASS PROPERTIES UNITS

Some mass properties are based on the geometric distribution of the object and are independent of the material's density (assumed to be uniform). Examples of these kinds of properties are volume and center of mass (centroid). Other properties, such as mass and moments of inertia, are dependent on the material's density. **Table 6-1** lists the densities of some common engineering materials. You can use these density values to set the material properties of your part. Weight and Mass are often confused. Below is an example of the correct way to calculate them for a unit one-inch cube. *Note*: The "Density" value used by SOLIDWORKS (see **Figure 6-5**) is actually the value of the "Unit Weight" listed below in **Table 6-1**. This lack of uniformity in terminology contributes to the general confusion on this matter.

Figure 6-5. Mass Property Options.

Example Calculation: Material is Mild Steel (density is 0.728×10^{-3} lbs-sec^2/in^4)

Weight   = (unit weight)x(volume) = (0.281 lbs/in^3) x (1.00 in^3) = 0.281 lbs.

Mass   = (density)x(volume) = (0.728×10^{-3} lbs-sec^2/in^4)x(1.00 in^3) = 0.728×10^{-3} lbs-sec^2/in.

          = (weight) / (gravity) = (0.281 lbs) / (386 in/sec^2) = 0.728×10^{-3} lbs-sec^2/in.

Table 6-1. Some Unit Weights and Densities of Common Materials.

Material	Unit Weight (lbs/in^3)	Density (lbs-sec^2/in^4)
Aluminum	0.097	0.251×10^{-3}
Brass	0.307	0.794×10^{-3}
Chromium	0.259	0.671×10^{-3}
Copper	0.323	0.837×10^{-3}
Magnesium	0.063	0.163×10^{-3}
Plastic	0.036	0.093×10^{-3}
Rubber	0.041	0.106×10^{-3}
Steel	0.281	0.728×10^{-3}
Titanium	0.163	0.422×10^{-3}

Exercise 6.1: ROCKER ARM MASS PROPERTIES

For this Exercise 6.1, you will build the basic geometry of the rocker arm, which is designed to rotate about a principal axis. You will then copy the first rocker arm data into a second rocker arm file and make a change in its geometry. Mass Properties analysis will be performed on both models and the results will be compared to each other.

ROCKER ARM 1

Open **ANSI-INCHES.prtdot** and **Save** it as **ROCKER ARM 1.sldprt**. In the **Front** plane, **Sketch** a **Circle** with a <u>diameter</u> of **1.75** inches and centered at the **ORIGIN**. **Extrude** it a **Blind** distance of **1.25** inches **OUT** from the front plane. You now have the bottom Boss/Base.

Return to the **Front** plane and **Sketch** the back upright feature of the Rocker Arm as indicated in **Figure 6-6**. Draw the three **Lines** such that the bottom of the profile slightly extends into the extruded base. Draw the outer top arc using the **Tangent Arc** tool. Draw the **Circle** below this arc. Now **Add Relations** to the profile as follows:

- o The center of the small circle and the origin are **Vertical**.
- o The small circle and top arc are **Concentric**.
- o The left-side line and top arc are **Tangent**.
- o The right-side line and top arc are **Tangent**.

Finally, use the **Dimension** tool to apply the three dimensions given in **Figure 6-6**. Now **Extrude** this upright sketch **0.75** inches **OUT** from the front plane. You will get the beginning model of the Rocker Arm as shown in **Figure 6-7**.

Now you will draw the final sketch to cut the hole and keyway through the bottom boss. Click on its front surface (it turns *blue*) and enter the **Sketch** mode. Set **Units** to inches and **3** decimal places. Use a **Front** view orientation to draw the big **Circle** with a <u>radius</u> of **0.50** inches and centered at the origin. Draw three short **Lines** to make the keyway at the top of the circle and then use the **Trim** tool to cut away the pieces that are not needed. Refer to **Figure 6-8** to see how it should look at this point. Use the **Dimension** tool to dimension the features in accordance with **Figure 6-8**.

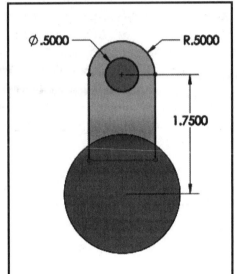

Figure 6-6. Dimensions for the Upright Part of the Rocker Arm.

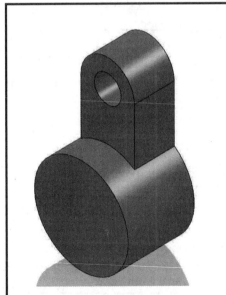

Figure 6-7. The Beginning Rocker Arm.

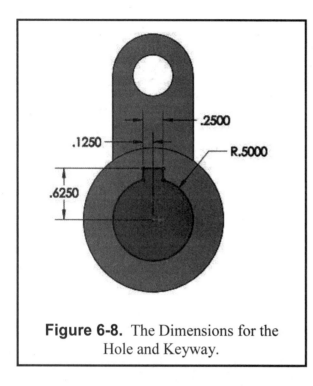

Figure 6-8. The Dimensions for the Hole and Keyway.

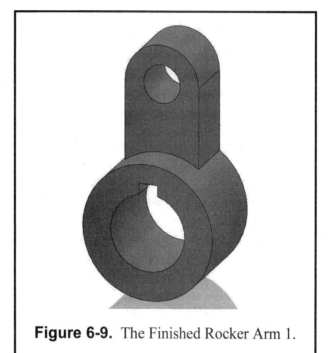

Figure 6-9. The Finished Rocker Arm 1.

Now **Extruded Cut** the sketch profile using a **Through All** end condition. The Rocker Arm 1 is finished as shown in **Figure 6-9** in a **Dimetric** view. Use the **Rotate View** icon to see the through hole and keyway features. **Save** your model as **ROCKER ARM 1**. Do not close your file.

ROCKER ARM 2

You will design Rocker Arm 2 by using the general database of Rocker Arm 1. So now **Save** your screen model as **ROCKER ARM 2.sldprt**. The three differences for the new Rocker Arm 2 are the width of the upright feature (now 1.25 inches) the height from the origin to the center of the circle (1.50 inches) and the depth of this upright feature (now only 0.50 inches). These changes can be accomplished easily using the feature editing capabilities of SOLIDWORKS.

In the Feature Manager tree, click over the **Sketch2** label with a right mouse click ("Sketch2" is under the second "Extrude" feature). Select the **Edit Sketch** option and the old "Sketch2" definition appears on the screen. Change the value of the large arc from R0.500 to **R0.625** and the height dimension from 1.75 to **1.500** as shown in **Figure 6-10**. Then click the **Rebuild** icon to execute this change in the upright features width.

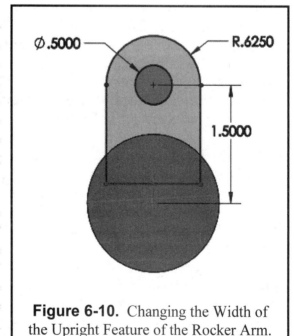

Figure 6-10. Changing the Width of the Upright Feature of the Rocker Arm.

Now change the extrusion depth of the upright feature. Right mouse click the second "Extrude" label in the Feature Manager tree. Select the **Edit Feature** option. The "Boss-Extrude" menu will now appear. In this menu, leave the end condition as **Blind**, but change "Distance" to **0.500** inches. Then click the green **OK** button. The solid model will now be rebuilt with the new extrusion depth. The Rocker Arm 2 is now finished as shown in **Figure 6-11** in a **Trimetric** view. Use the **Rotate View** icon to see the through hole and keyway features. **Save** your model as **ROCKER ARM 2**. Then **Close** your SOLIDWORKS file to start the next phase of this analysis exercise.

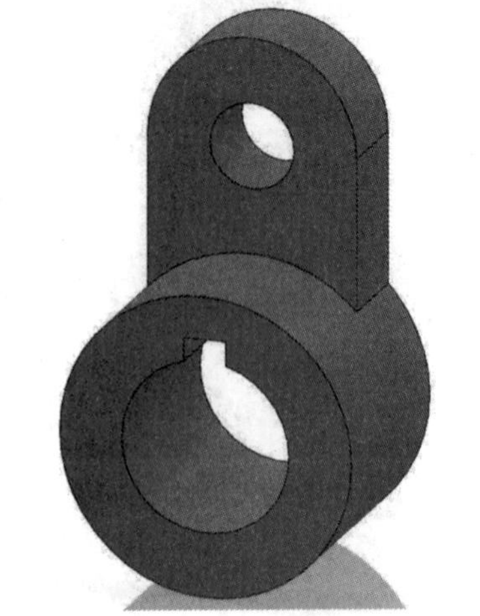

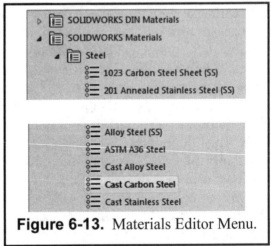

Figure 6-11. The Finished Rocker Arm 2.

MASS PROPERTIES ANALYSIS OF ROCKER ARM 1

You will start your mass properties analysis with Rocker Arm 1. Pull down **File**, select **Open**, and open the **ROCKER ARM 1** file from your folder. Under **Tools – Options – Document Properties – Units - Mass/Section Properties** change **Length** to **4 decimal places**. Now you need to set the material for the Rocker Arm 1. Right click on the **Material** icon (**Figure 6-12**) in the Feature Manager and select **Edit Material**. Under **SOLIDWORKS Materials** in the "Materials Editor" menu (**Figure 6-13**), expand the **Steel** library of materials and select **CAST CARBON STEEL**. If you do not want the part color to change with material assignment, then **Click** on the **Appearance Tab** and click **off** the **"Apply Appearance of: Cast Carbon Steel"** box. Next, select **Apply** and then **Close.**

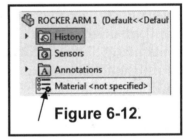

Figure 6-12.

Pull down **Tools, Evaluate**, and select the **Mass Properties** command. The "Mass Properties" report window will now appear on the screen showing all the mass properties for Rocker Arm 1 for cast carbon steel. If you cannot see all of the report, pull down on the lower right corner of the

Figure 6-13. Materials Editor Menu.

window border to stretch the "Mass Properties" window as shown in **Figure 6-14**. Recall the options of the "Mass Properties" buttons, which includes the ability to print this report to a local printer. However, you may prefer to place your name and other data on the hard copy report. So, in the "Mass Properties" window, click the **Copy** button. This will copy the entire report to the computer's clipboard. Now open some word processing software like MS Word. Use the **Edit** and **Paste** commands in the software to paste the clipboard data directly onto a new page. At the top of the Mass Properties report, type in Your **Name,** Your **Seat** Number, and Your

Section Number. Also add an opening sentence of the Mass Properties report: "**Mass Properties of Rocker Arm 1 with Cast Carbon Steel Material**" in order to identify the part name and assigned material. Next, select **File** and **Save** in the word processing software to save the word file as **ROCKER ARM 1 - STEEL**. Then close your word processing software and return to SOLIDWORKS. You will print this document after you determine which of the mass property reports satisfies the question at the bottom of **Page 6-8**.

When you return to SOLIDWORKS, the "Mass Properties" window is still open, so click the **Close** button. You will now repeat this mass properties analysis for Rocker Arm 1 using **Aluminum** as the material.

Right click on the **Material** (currently Cast Carbon Steel) icon (**Figure 6-12**) in the Feature Manager. Under **SOLIDWORKS Materials**, expand the **Aluminum** library of materials and select **1060 ALLOY.** This sets all the properties for this item. If you do not want the part color to change with material assignment, then **Click on the Appearance Tab** and click **off** the "**Apply appearance of: 1060 ALLOY**" box.

Pull down **Tools, Evaluate**, and select the **Mass Properties** command. The "Mass Properties" report window will now appear on the screen showing all the mass properties for Rocker Arm 1 with aluminum material, as shown in **Figure 6-15**. Now repeat the **Copy** to clipboard and **Paste** in a word processing software procedure used previously. Be sure to include your pertinent data and material on the file. **Save** your **ROCKER ARM 1 - ALUMINUM** file and then you can close the word processing software. You will print this document after you determine which of the mass property reports satisfies the question at the bottom of **Page 6-8**.

When you return to SOLIDWORKS, **Close** your "Mass Properties" window. You can then **File, Close** your **ROCKER ARM 1** part file to start analysis on Rocker Arm 2.

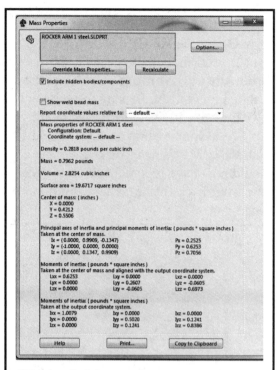

Figure 6-14. Mass Properties for Cast Carbon Steel Rocker Arm 1.

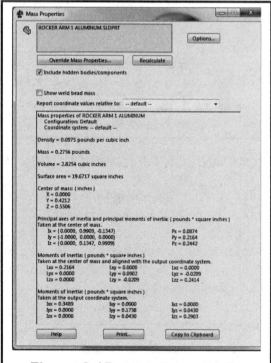

Figure 6-15. Mass Properties for Aluminum Rocker Arm 1.

MASS PROPERTIES ANALYSIS OF ROCKER ARM 2

Now **Open** the **ROCKER ARM 2** file and repeat this mass properties analysis with the Rocker Arm 2 solid model. Repeat all the previous steps, first for **Cast Carbon Steel** and then for **1060 Alloy Aluminum**. Get word files of these two reports using the copy and paste procedure outlined previously. Be sure to type your name, class data and material on the reports before you **SAVE** them. Refer to **Figures 6-16** and **6-17** as needed for these mass property reports. You can **Save** your reports as files named **ROCKER ARM 2 - STEEL** and **ROCKER ARM 2 - ALUMINUM**. When you are finished, **Close** down your **ROCKER ARM 2** part file. You should now have reports of various mass property reports saved. You will print these documents after you determine which of the mass property reports satisfies the question at the bottom of **Page 6-8**.

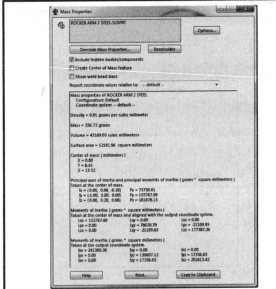

Figure 6-16. Mass Properties for Cast Carbon Steel Rocker Arm 2.

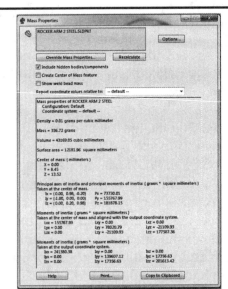

Figure 6-17. Mass Properties for Aluminum Rocker Arm 2.

COMPARISON OF MASS PROPERTIES FOR ROCKER ARMS 1 and 2

After you have completed the mass properties analysis for both Rocker Arms 1 and 2, you can make some comparisons. Because of the design and functionality of the rocker arm, a key parameter is motion about the central axis of the large cylindrical hole on the bottom. In this case, it would be motion about the Z-axis. Hence, the moment of inertia (resistance to steady rotation) about the Z-axis taken at this center would be considered a critical physical property. Study the moment of inertia about the Z-axis of the two rocker arm designs for each of the two assigned materials (mild steel and aluminum), and answer the question below on the correct Mass Property report.

> **Note:** You now have four Mass Properties reports. Identify on the correct report, which combination of geometry and material yields a Rocker Arm design that is easiest to rotate (requires the least torque) about the central Z-axis?

CREATING A SHADED IMAGE OF THE TWO ROCKER ARMS

You will now create shaded images of the two Rocker Arms, side-by-side, and obtain a plot for submission to your instructor. You will do this using an assembly model. Pull down **File**, select **Open**, and then select the **ROCKER ARM 2** part file. Repeat this **File**, **Open** sequence for **ROCKER ARM 1** also. Now pull down **File** and select **New - Assembly**. This will create a blank assembly file. Pull down **File**, select **Save As**, and then give your new assembly file the name **ROCKER ARM ASSEMBLY**. At first, all of these computer screen images (viewports) land on top of each other. To see all three screens, pull down **Window** and select **Tile Vertically**. This will result in a screen image as shown in **Figure 6-18**.

To assemble your two Rocker Arms, just pick, drag, and drop your parts into the assembly window. Click on the **ROCKER ARM 1** icon in its Feature Manager window. Drag it into the assembly window and drop it into an area around the origin. In like manner, go to the **ROCKER ARM 2** icon in its Feature Manager window and drag and drop it into an area just to the right of Rocker Arm 1. Both parts are now in the assembly. You no longer need the two Rocker Arm files, so you can close both of those windows by clicking in the box ☒ symbol at the upper right corner of their window borders. Also, you can now stretch or maximize your Rocker Assembly window by clicking the maximize box in the upper right corner of its window border.

The Rocker Arms may have been dropped into somewhat arbitrary positions. So first try a **Trimetric** View Orientation. Now arrange Rocker Arm 2 to be on the right side of Rocker Arm 1. Select the **Move Component** icon (refer back to **Figure 5-3**) from the Assembly Toolbar. With the move cursor, pick Rocker Arm 2 and move it to an acceptable position (see **Figure 6-19** as a guide). You can also use **Zoom** and **Pan** to arrange your image. You may now save your assembly, so pull down **File** and select **Save As**. Save your **ROCKER ARM ASSEMBLY**.

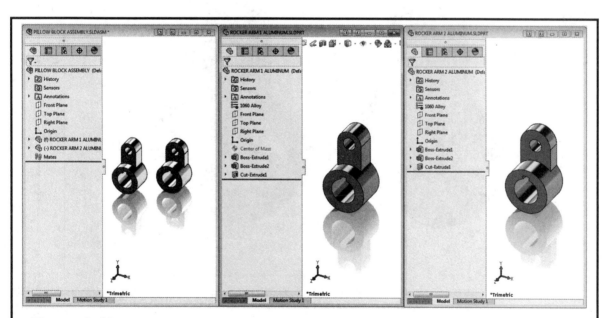

Figure 6-18. The Rocker Arm Assembly, Rocker Arm 1, and Rocker Arm 2 Windows Tiled Vertically.

Open the **TitleBlock-Inches.drwdot** from your folder and insert a dimetric view of the Assembly in the same manner that you have been doing with single solid parts. Before you print a hardcopy of your image, you need to label the Rocker Arms. Pull down the **Insert** menu, select **Annotations**, and then select **Note**. The on-screen "Note" menu appears. Type in the label "**ROCKER ARM 1**" in the note window. You may want to dictate the font size also, so click **off** the check mark in the "Use Documents Font window." Click the **Font** button and the "Choose Font" on-screen menu appears. Here you can set the font type and height. Use the **Arial** font type with a **Bold** setting and set the height to **0.20** inches. Then click **OK** twice to close the on-screen menus, and the note appears on the screen. In all likelihood, the note is not positioned correctly, so drag it with the LMB to an appropriate position above **ROCKER ARM 1**. Repeat this annotation process for **ROCKER ARM 2**. Refer to **Figure 6-19** for an example of an acceptable screen layout.

Now preview your print file by pulling down **File** and select **Print Preview**. See your image as it will appear when printed as shown in **Figure 6-19**, and then click the **Close** button. Then pull down **File** and select **Print**. Make sure the proper printer is selected, and then click **OK**. This finishes the Lab Exercise 6.1 and you can pull down **File** and click **Close** to exit.

Note: **Attach your Rocker Arm Assembly printout to your four Mass Properties reports and submit them to your instructor before you leave the lab**.

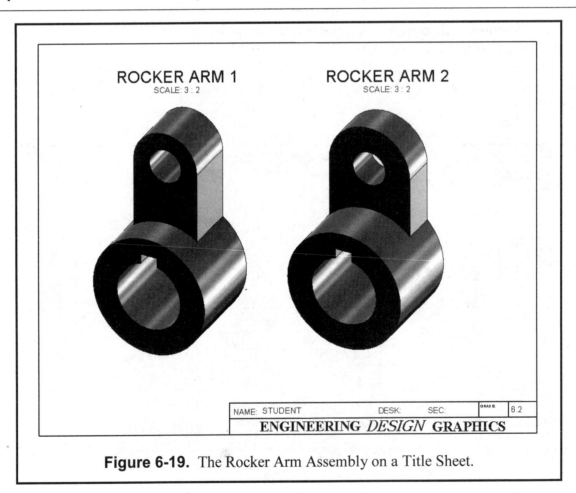

Figure 6-19. The Rocker Arm Assembly on a Title Sheet.

INFORMATION PROVIDED

IN

SOLIDWORKS MASS PROPERTIES REPORTS

1. **DENSITY** is the mass or the weight per unit volume of the material the part is made from.

2. **MASS**: The mass of a body is the measure of its property to resist change in its steady motion. The mass depends on the volume of the body and the density of the material of which the body is made. In this case with SOLIDWORKS, mass is equivalent to weight.

3. **VOLUME**: The volume of a body is the total volume of space enclosed by its boundary surfaces.

4. **SURFACE AREA**: The surface area is the total area of the boundary surfaces defining the solid model.

5. **CENTER OF MASS**: Center of Mass (or Centroid) of a volume is the origin of coordinate axes for which first moments of the volume are zero. It is considered the center of a volume. For a homogeneous body in a parallel gravity field, mass center and center of gravity coincide with the centroid.

6. **PRINCIPAL AXES OF INERTIA AND PRINCIPAL MOMENTS OF INERTIA**: Principal moments of inertia are extreme (maximal, minimal) moments of inertia for a body. They are associated with principal axes of inertia which have its origin at the centroid and the direction of each usually given by the three unit-vector components. For these axes, the products of inertia are zero.

7. **MOMENTS OF INERTIA**: A moment of inertia is the second moment of mass of a body relative to an axis, usually X, Y, or Z. It is a measure of the body's property to resist change in its steady rotation about that axis. It depends on the body's mass and its distribution around the axis of interest.

Exercise 6.2: SOCKET DESIGN TABLE

In Exercise 6.2, you will make a family of parts of the Socket using a Design Table. A Design Table allows you to build multiple configurations of parts or assemblies by specifying parameters in an embedded Microsoft Excel worksheet. The parameters can be the dimensions of key features of the design object that will vary from one part configuration to the next. Once the design table is built, you can go to the "Configuration Manager" to sequence through the different parts. You can then load all your configured parts into an assembly image to display them.

Open **ANSI-INCHES.prtdot** and **Save** it as **SOCKET.sldprt**. Start a **Sketch** in the **Front** plane. Use the **Centerline** and **Rectangle** tools to create the sketch profile shown in **Figure 6-20**. Make sure the vertical centerline goes through the origin, and that the upper left corner of the rectangle is also at the origin. Now carefully **Dimension** the sketch profile **in this order.**

First, the bottom **.3850** overall width.
Second, the **.9600** inch overall height.

While you have been applying and fixing these dimension values, SOLIDWORKS has also been keeping track of them with a more general approach. It names each dimension with a standard variable name "D#@Sketch#" where the first # is the order the dimension was applied to the sketch and the second # is the Sketch number. For example, the first .3850 inch dimension should be labeled "D1@Sketch1." This information is critical in the creation of a Design Table in SOLIDWORKS, since you can assign different values to this general dimension variable.

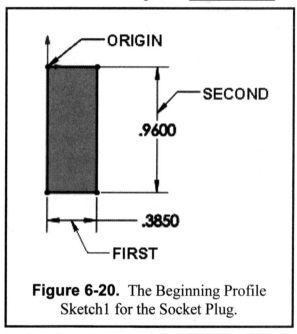

Figure 6-20. The Beginning Profile Sketch1 for the Socket Plug.

Now explore this dimension numbering scheme and revise your sketch accordingly. In the **Select** mode, position your mouse over the first .3850 dimension and select it (**LMB**). A Dimension menu appears over the Feature Manager Tree (shown in **Figure 6-21**). Notice the window on the menu that is titled "Primary Value." It lists the software name for the dimension as **D1@Sketch1** and gives the value of that dimension (**.3850 in**). The window below the **Primary Value** is labeled "**Dimension Text.**" In this window you will see <DIM>. With one space between, type the dimension number with the parentheses **(D1)**. **Make sure this number corresponds to the dimension name in the Primary Value window.** Observe the "Preview" of this label **.3850 (D1)** on your sketch. This is just a reference label to help you later identify it in your Design Table.

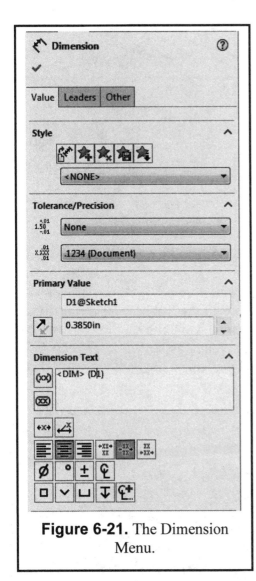

Figure 6-21. The Dimension Menu.

Now using this same procedure, add the corresponding (D#) label to the remaining dimension in the same order you applied it earlier. In each case, use the following same steps:

LMB on top of the dimension
The "Dimension" Menu appears
Observe its "Full Name"
Click in the **Dimension Text** Box
Type in the proper **(D#)**
Observe the "Preview"
Click **OK**.

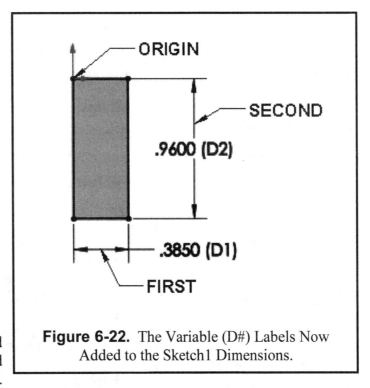

Figure 6-22. The Variable (D#) Labels Now Added to the Sketch1 Dimensions.

When you are finished, your sketch and corresponding dimensions should appear as shown in **Figure 6-22**. Study these to make sure they correspond to the order in which they were applied. Now click the **Base Revolve** icon and revolve the sketch **360** degrees to create the base part for the Socket, as shown in **Figure 6-23**. Activate the **Features Tab** and select the **Fillet** command. Add a .**03 inch** fillet to the top edge of the **Base Part** as shown in **Figure 6-24**.

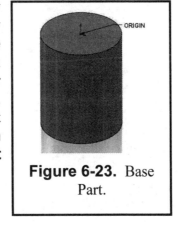

Figure 6-23. Base Part.

Figure 6-24. Base Part with Fillet.

Now you will create the square hole in the bottom of the socket where the ratchet wrench fits. Click on the **Bottom** surface on the Base Part; also select a **Bottom** view orientation. In the **Sketch** mode, draw a **Rectangle** as shown in **Figure 6-25**. Use the **Dimension** tool to apply the nominal dimension values indicated in **Figure 6-25**. Next, execute an **Extruded Cut** for a distance of **.4500 inches**.

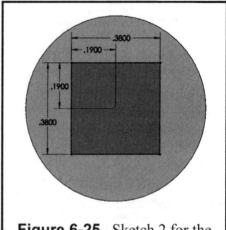

Figure 6-25. Sketch 2 for the Ratchet Wrench Opening.

Select the top surface of the Socket. Activate the Sketch command and Sketch a **hexagon** with an **inscribed circle** diameter of **.5660 in.** at **0 degrees (Figure 6-26 and 6-27)**. Even though it is easy to insert the hexagon with a diameter of .5660 in the feature manager tree, it is not a dimension that you can use in the design table later on in this exercise. You need a dimension in the sketch for use in the Table. Select the **Smart Dimension** to dimension the inscribed circle to **.5660 inches**. Now use the previous steps to **Label** the dimension so that its label **(D#)** appears with its value, as shown in **Figure 6-27**. **Extrude Cut** this pattern **Up To Surface** and then **select** the **top** of the **square hole** on the bottom of the Socket.

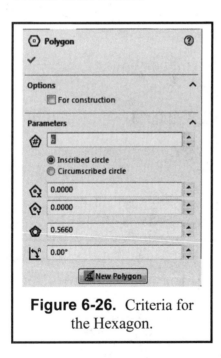

Figure 6-26. Criteria for the Hexagon.

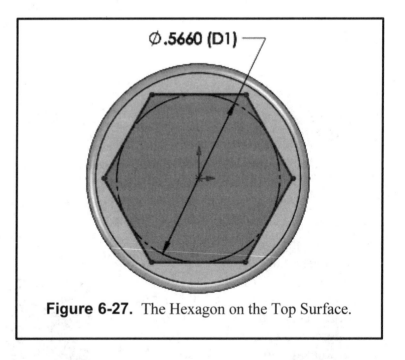

Figure 6-27. The Hexagon on the Top Surface.

Select the top surface of the Socket again. Activate the Sketch command and repeat the whole process of the previous hexagon with all of the same dimensions except for the angle of the hexagon should be set to **30 degrees**. You need a dimension in the sketch for use in the table. Select the **Smart Dimension** to dimension the inscribed circle to **.5660 inches**. Now use the previous steps to **Label** the dimension so that its label **(D#)** appears with its value, as shown in **Figure 6-29**. See **Figure 6-28 & 6-29**. **Extrude Cut** this hexagon **Up To Surface** which is the bottom of the first hexagonal cut.

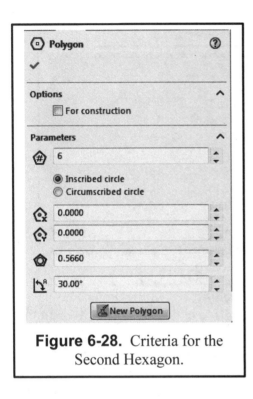

Figure 6-28. Criteria for the Second Hexagon.

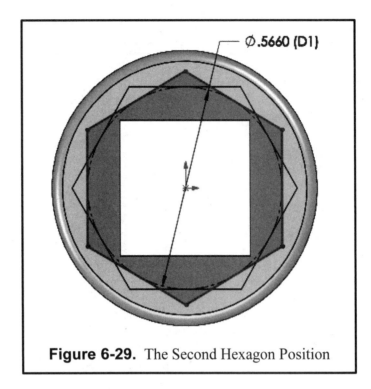

Figure 6-29. The Second Hexagon Position

The last step in the construction of the Socket is to **Chamfer** all of the short edges of the polygon on the top surface of the Socket. Set the value of the **Chamfer** to **.023 in**. The finished socket should look like the socket in **Figure 6-30**. Save your model as **SOCKET.sldprt**.

You are ready to make a Design Table. Pull down **Insert**, select **Tables – Design Table**, and then select **Blank** on the Design Table menu. Click on the green check mark and a blank Excel spreadsheet will appear on your screen as suggested in **Figure 6-31**. There is some preliminary default data on the spreadsheet. Row 1 will have a label stating "Design Table for: Socket" and Row 3, column A will have "First Instance." Key in the following new configuration titles and dimension variables in the Excel editing box (in the **top toolbar area**):

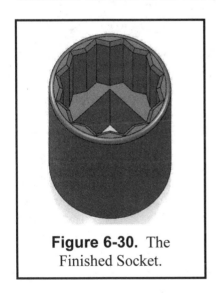

Figure 6-30. The Finished Socket.

Row 3, Column A = **First Instance**
Row 4, Column A = **Second Instance**
Row 5, Column A = **Third Instance**
Row 6, Column A = **Fourth Instance**
Row 7, Column A = **Fifth Instance**
Row 8, Column A = **Sixth Instance**

Row 2, Column B = **D1@Sketch1**
Row 2, Column C = **D2@Sketch1**
Row 2, Column D = **D1@Sketch3**
Row 2, Column E = **D1@Sketch4**

Make sure the sketch numbers in your table correspond to <u>your</u> sketch numbers

	A	B	C	D	E	F
1	Design Table for: SOCKET					
2		D1@SKETCH1	D2@SKETCH1	D1@SKETCH3	D1@SKETCH4	
3	First Instance	0.385	0.96	0.566	0.566	
4	Second Instance	0.432	1.005	0.629	0.629	
5	Third Instance	0.474	1.018	0.692	0.692	
6	Fourth Instance	0.502	1.132	0.755	0.755	
7	Fifith Instance	0.542	1.12	0.818	0.818	
8	Sixth Instance	0.584	1.255	0.88	0.88	
9						

Sheet1

Figure 6-31. The Design Table for the Six Instances of the Socket.

Now study the data in the Design Table in **Figure 6-31**. Carefully key in the dimension values for D1@Sketch1 for Configurations 1 through 6 in their proper place under column B. Repeat this data entry for D2@Sketch1 under column C, D1@Sketch3 under column D, and D1@Sketch4 under column E. <u>*Note:*</u> **Be careful to keep your cursor <u>inside</u> the spreadsheet's confines, because when you click outside of it, the Design Table is executed and the spreadsheet disappears.** So when you are finished entering the data, just click on the computer screen to launch the Design Table. If you need to re-access your design table, select the Configuration Manager Icon **(Figure 6- 32)**, click the + sign in front of Tables in the Feature Manager Tree and right click on Design Table. Then select Edit Table.

If you have done this correctly, a Design Table icon is added to the Feature Manager tree, as shown in **Figure 6-32**. From here you can Right Mouse click (**RMB**) and **Edit Table** of the Design Table to get it back on the screen. <u>*Note:*</u> Sometimes when you are completing a Design Table, you may make a mistake or inadvertently close it before you are finished. Right click on the **Design Table** icon in the **Feature Manager** and **Edit Table** to get it back on the screen (See **Figure 6.34**). Even if you believe you entered the data correctly, it might be a good idea to <u>check it again</u> before going forward.

Figure 6-32. The Configuration Manager Icon.

You are now ready to see the six different design instances created by the Design Table. First click the **Rebuild** icon (green light symbol) to rebuild all your design data. You might also want to **Save** your part file now to be safe.

Notice the small instance icon at the top of the Feature manager column, as indicated in **Figure 6-32**. Click this **Configuration Manager** icon to turn the Feature

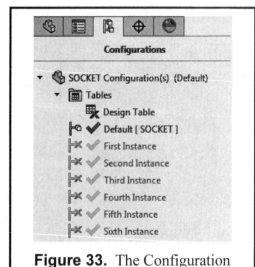

Figure 33. The Configuration Manager Tree.

Manager area into a "Configuration Manager" tree, as shown in **Figure 6-33**.

Now in an **Isometric** view, one-by-one, double click on the grayed **First Instance**, **Second Instance**, **Third Instance**, **ETC. labels** in this tree to see the six different configurations of the Socket. You may have to **Rebuild** your data during this process (See **Figure 6-34**). Also click on the **Default** configuration, which also happens to be the same as "First Instance" because the values entered in row 3 just happen to be exactly the same as your original sketch data. *Note*: First Instance does not have to be the same as the "Default" configuration. **Save** your file again.

Since this is a family of parts, it would be nice to view them all together to see their variations in size and shape. One way to do this is to "drag and drop" them all into an assembly. So pull down **File**, **New** and select **Assembly**. Pull down **Window** and **Tile Vertically**. Now "drag and drop" the configurations one-by-one, into the assembly window. Start by selecting the "First Instance" label in the Configuration Manager tree, then "Second Instance," then Third Instance," and through the "Sixth Instance." You can now maximize the assembly window to see the operation better. Arrange them in the assembly window using the **Move Component** icon tool to get a layout as suggested in **Figure 6-35** without the labels. Once aligned, save as an assembly. Open **TITLEBLOCK-INCHES** and insert

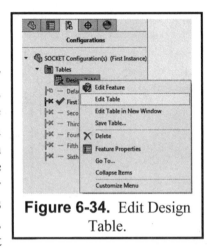

Figure 6-34. Edit Design Table.

your assembly drawing in the same way you have been inserting your solid models. Add **Insert, Annotation** titles and **Print** the image to hand in to your instructor before you leave the lab. **Save** your file as **SOCKET ASSEMBLY.slddrw** and then log off.

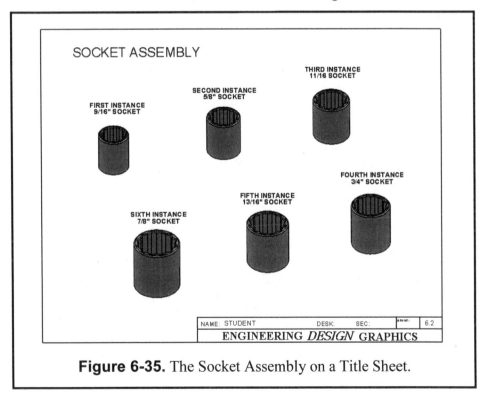

Figure 6-35. The Socket Assembly on a Title Sheet.

SUPPLEMENTARY EXERCISE 6-1: CASTER AND CASTER FRAME DESIGN TABLES

Create a Design Table to generate the series of Caster Frames and/or Caster Wheels as specified in the Table provided. See your instructor for specific instructions.

1. Generate an assembly drawing of the six **Caster Frames** produced from the Design Table and insert it into a Title Block (see **Figure 6-30** as an example).

2. Generate an assembly drawing of the six **Caster Wheels** produced from the Design Table and insert it into a Title Block (see **Figure 6-30** as an example).

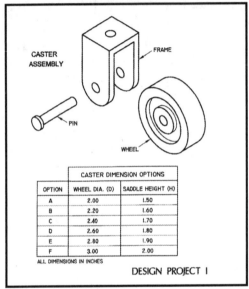

CASTER DIMENSION OPTIONS		
OPTION	WHEEL DIA. (D)	SADDLE HEIGHT (H)
A	2.00	1.50
B	2.20	1.60
C	2.40	1.70
D	2.60	1.80
E	2.80	1.90
F	3.00	2.00

ALL DIMENSIONS IN INCHES

DESIGN PROJECT 1

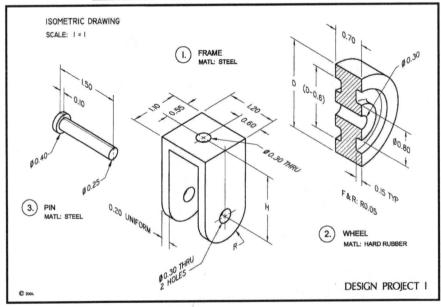

DESIGN PROJECT 1

SUPPLEMENTARY EXERCISE 6-2: PULLEY AND PULLEY BRACKET DESIGN TABLES

Create a Design Table to generate the series of Caster Frames and/or Caster Wheels as specified in the Table provided. See your instructor for specific instructions.

1. Generate an assembly drawing of the six **Pulley Brackets** produced from the Design Table and insert it into a Title Block (see **Figure 6-30** as an example).

2. Generate an assembly drawing of the six **Pulleys** produced from the Design Table and insert it into a Title Block (see **Figure 6-30** as an example).

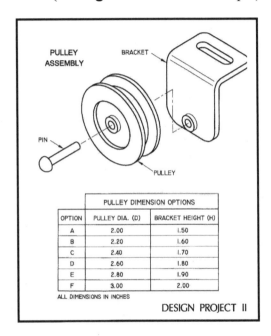

PULLEY DIMENSION OPTIONS		
OPTION	PULLEY DIA. (D)	BRACKET HEIGHT (H)
A	2.00	1.50
B	2.20	1.60
C	2.40	1.70
D	2.60	1.80
E	2.80	1.90
F	3.00	2.00

ALL DIMENSIONS IN INCHES

DESIGN PROJECT II

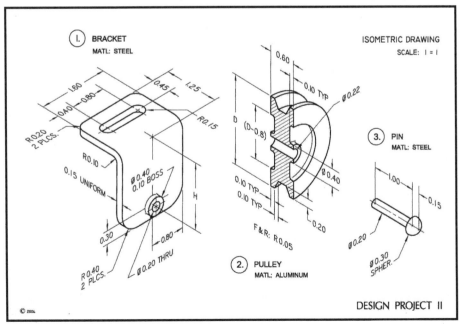

DESIGN PROJECT II

NOTES:

Computer Graphics Lab 7:
Analysis and Design Modification II

INTRODUCTION

SOLIDWORKS parametric modeling software allows the designer to build and modify 3-D solid models easily by editing the dimensions and constraints imposed on the 2-D sketch, and then rebuilding the model. In this lab exercise, you will build a 3-D model of a pillow block model using an initial design concept. The initial 3-D model database will then be analyzed using the **SOLIDWORKS Simulation** add-in. You will perform a finite element analysis (FEA) on the pillow block to test the stress in the part during its primary function, which is to support a rotating shaft. The FEA results will then suggest modifications to the initial design of the pillow block in order to improve its strength and performance. You will then redesign the pillow block using the parametric features of SOLIDWORKS. Once the second design concept is completed, you will make a second study of the pillow block to see if the design changes improved the performance of the pillow block.

Procedure for Finite Element Analysis using SOLIDWORKS Simulation

The purpose of this exercise is to learn the stage in the design process subsequent to modeling the object. This is the Finite Element Analysis of the solid model. The results of this analysis are used to evaluate the design of the component. Finite Element Analysis allows the user to simulate a variety of load conditions that the part would be subjected to in the environment that it is designed for and evaluate the performance of the part under them before physically building the part. It is also extremely useful when destructive testing of the part in question is prohibitively expensive. SOLIDWORKS Simulation in conjunction with SOLIDWORKS offers a powerful tool for FEA. The part can be built as a solid model and SOLIDWORKS Simulation can be used to analyze the performance of the part.

The object of this exercise is to build a simple part (A Pillow Block) in SOLIDWORKS and analyze it using SOLIDWORKS Simulation. Before beginning the exercise SOLIDWORKS Simulation should be installed as an "Add In" to SOLIDWORKS. The steps for accomplishing this are outlined below:

1. Open SOLIDWORKS
2. Click on **TOOLS** - **Add-Ins** and select **SOLIDWORKS Simulation**

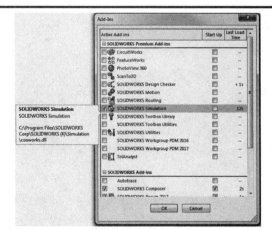

Figure 7-1. SOLIDWORKS Simulation
Add-In.

This adds a Simulation Tab at the top of the work area. Right click in the gray area at the top of the screen and choose the Simulation Toolbar. The menu and the tools are displayed when a new file is created or an existing file is opened. The Simulation Tool Bar will have the following icons. The name of each of the icons is given in **Figure 7-2** and a written description of each of the commands is given in **Table 7-1**.

Table 1. Definitions

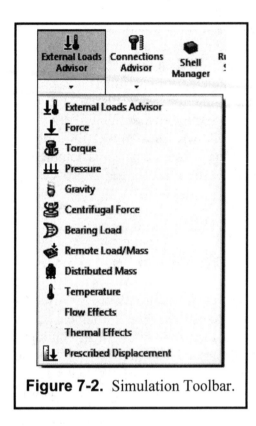

Figure 7-2. Simulation Toolbar.

Restraints	Applies restraints to the selected entities for the active structural study (static, frequency, or buckling).
Pressure	Applies pressure to the selected faces for the active structural study (static, frequency, or buckling).
Force	Applies force, torque, or moment to the selected entities for the active structural study (static, frequency, or buckling). The specified value is applied to each selected entity.
Gravity	Defines gravity loading for the active structural study (static, frequency, or buckling).
Centripetal Force	Applies centripetal forces for the active structural study (static, frequency, or buckling).
Remote Load	Applies remote loads for structural studies.
Rigid Connection	Applies rigid connection between selected faces in structural studies.
Bearing Load	Applies bearing loads on selected faces of different components.
Temperature	Applies temperatures on the selected entities for the active thermal or structural (static, frequency, or buckling) studies.

Exercise 7.1: FINITE ELEMENT ANALYSIS OF A PILLOW BLOCK

DESIGNING THE FIRST VERSION OF THE PILLOW BLOCK

You will start by building the preliminary Pillow Block. Study the needed geometry for this version in **Figure 7-3**. To start, open **ANSI-INCHES.prtdot** from your folder. Pull down **File** again, select **Save As**, key in the name **PILLOW-1.sldprt**, make sure the directory is set to your folder, and then click the **Save** button. Notice that the part name is now "Pillow1" in the Feature Manager tree structure on the screen.

Select the **Front** plane in the Feature Manager. Enter the sketch mode by clicking the **Sketch Mode** icon on the Sketch Toolbars and reproduce the front profile with the center of the large hole at the **ORIGIN** and then extrude the profile about **Mid-Plane** to the depth of **1.40** inches (See **Figure 7-4**). In the **Feature Manager**, select the **Top Plane**. **Extrude-Cut** the through holes in the base according to the dimensions given in **Figure 7-5**.

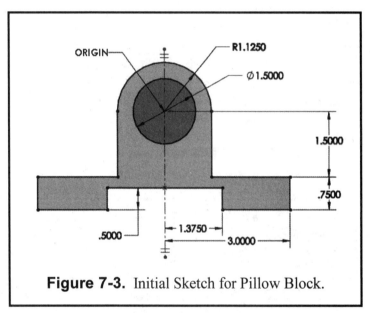

Figure 7-3. Initial Sketch for Pillow Block.

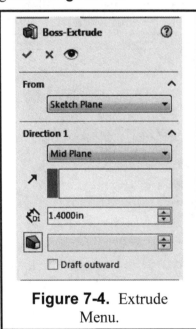

Figure 7-4. Extrude Menu.

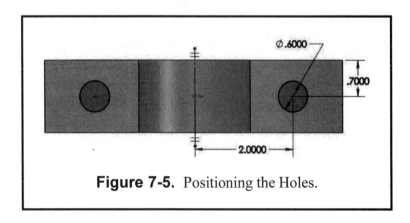

Figure 7-5. Positioning the Holes.

When finished, your solid model should look like the Pillow Block in **Figure 7-6**. **Save PILLOW1.sldprt** to your folder.

Now you will build a shaft that fits into the large hole of the Pillow Block. To start, open **ANSI-INCHES.prtdot**. Pull down **File** again, select **Save As**, key in the name **SHAFT.sldprt**, make sure the directory is set to your folder, and then click the **Save** button. Notice that the part name is now Shaft in the Feature Manager tree structure on the screen.

Select the **Front** plane in the Feature Manager. Click the **Sketch Mode** icon on the Sketch Toolbars and draw a **Circle** with a diameter of **1.50** inches at the origin (See **Figure 7-7**) and then extrude the profile about the **Mid-Plane** to the depth of **7.00** inches (See **Figure 7-8**). Add a **Chamfer** of **0.125** (**Figure 7-9**) on both ends of the shaft to complete this solid model. See **Figure 7-10**.

Save SHAFT.sldprt to your folder.

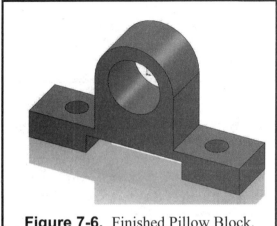

Figure 7-6. Finished Pillow Block.

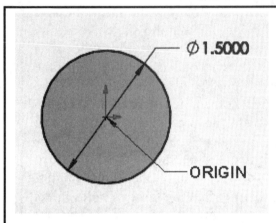

Figure 7-7. Circle Sketch for Shaft.

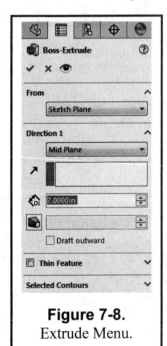

Figure 7-8. Extrude Menu.

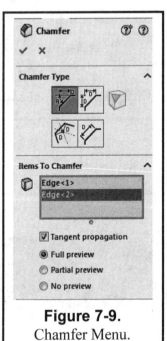

Figure 7-9. Chamfer Menu.

Figure 7-10. The Finished Shaft.

ASSEMBLY OF PARTS

At this time you are to make an assembly of the two parts. Make sure that you have Pillow1 and Shaft open. Go to **File**, **New** and select **Assembly**; set the **Units** to **MMGS**. Go to the pull-down **Tools** menu and select **Add-Ins**. Check **SOLIDWORKS Simulation** from the available list. Next, pull-down **View** and turn on **Origins**. **Save As "PILLOW ASSEMBLY-1.SLDASM."** Go to the **Window** pull-down menu and select "**Tile Vertically**." First, **Select** the Pillow Block and drag it to the **Origin (as your cursor approaches the origin, you will see a double set of arrows)** of the Assembly drawing. Next, **Select** the **Shaft** and drag it to the **Origin (as your cursor approaches the origin, you will see a double set of arrows)** of the Assembly Drawing. This will center the shaft inside of the hole of the pillow block. Maximize the Assembly file Window and view the assembly in isometric (See **Figure 7-11**).

Figure 7-11. Pillow Block and Shaft

FINITE ELEMENT ANALYSIS USING SOLIDWORKS SIMULATION

The actual analysis of the pillow block could be a complex procedure even with SOLIDWORKS Simulation, since analysis of the pillow block would have to include the effects of the rotating shaft and thermal stress on the shaft and the pillow block. In this case only a simple static analysis of the pillow block will be carried out.

If the Simulation Tab does not appear in the menu, go to **Tools**, **Add-ins** and select **SOLIDWORKS Simulation** and select **OK**. This will place the pull down tab in the menu and the top of your screen.

To begin analyzing the part, a study needs to be defined. The steps required for defining a study are:

Figure 7-12. Simulation Tab.

1. Click on the **SIMULATION TAB** at the top of the screen (**Figure 7-12**).
2. Select **New Study - New Study (Figure 7-13)**.
3. Click in the box below **Study, Name**, and name the study (e.g. **Study-1- with your last name**) as shown in **Figure 7-14**, and select the type of study as **STATIC**. Then click **OK**.
4. The Study Manager Tree appears at the lower left of the screen. See **Figure 7-15**.

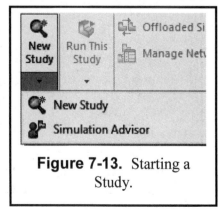

Figure 7-13. Starting a Study.

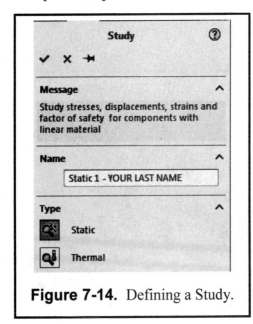

Figure 7-14. Defining a Study.

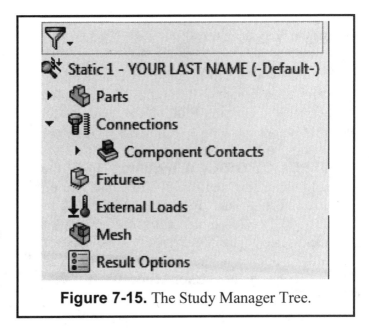

Figure 7-15. The Study Manager Tree.

Assigning Material to the Parts

SOLIDWORKS Simulation has a large material library, which contains the properties of a wide variety of materials. It also allows the user to create a new material with properties defined by the user. The procedure to apply a material is outlined below:

1. Go to the Study Manager Tree and click on the small arrow next to "Parts" to expand the **Parts** Feature. Right click on **Pillow1 (Figure 7-15)**, and select the **Apply/Edit Material** option.
2. In the Material window (**Figure 7-16**) select **From Library File – SOLIDWORKS materials.** Assign **Cast Alloy Steel** to the **Pillow Block.** Verify that the "**Type**" is set at **Linear Elastic Isotropic** and that the "**Units**" are set to **SI – N/mm^2 (MPa)** and click on **OK**. This assigns the material to the **Pillow Block.**
3. Repeat this process for the **Shaft** using **Alloy Steel** as the material.

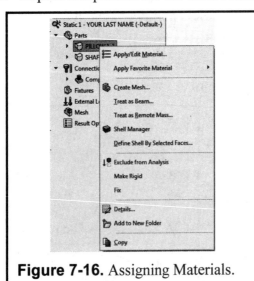

Figure 7-16. Assigning Materials.

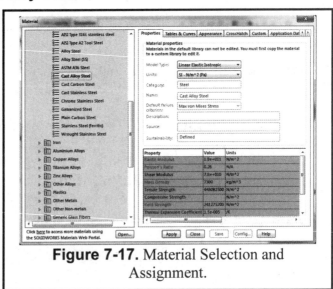

Figure 7-17. Material Selection and Assignment.

4. A *green check mark* appears on the icon next to the name of the model under the Solids folder in the SOLIDWORKS Simulation Feature Manager indicating that a material has been defined.

Applying Restraints:

A restraint is placed on the bottom surfaces of the pillow block to fix the object for a simple static analysis. The sequence of steps for this is follows:

1. Right click on **Fixtures** and select **Fixed Geometry.**
2. In the resulting window you are to select the two bottom faces of the Pillow Block as illustrated in **Figure 7-19**.

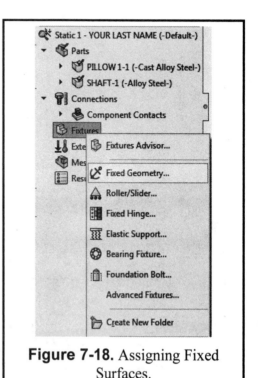

Figure 7-18. Assigning Fixed Surfaces.

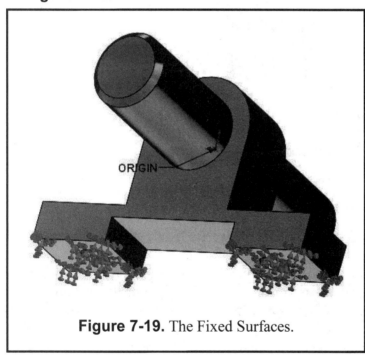

Figure 7-19. The Fixed Surfaces.

Applying the Force on the Shaft

In the static analysis the Pillow Block will react to the force the shaft exerts. The sequence of steps to apply this force is described below:

1. Right click on the **External Loads** icon in the SOLIDWORKS Simulation Feature Manager tree and select **Force** (see **Figure 7-20**).
2. For the **First Selection** box, click on the cylindrical surface of the shaft.
3. Next, choose **"Selected Direction."**

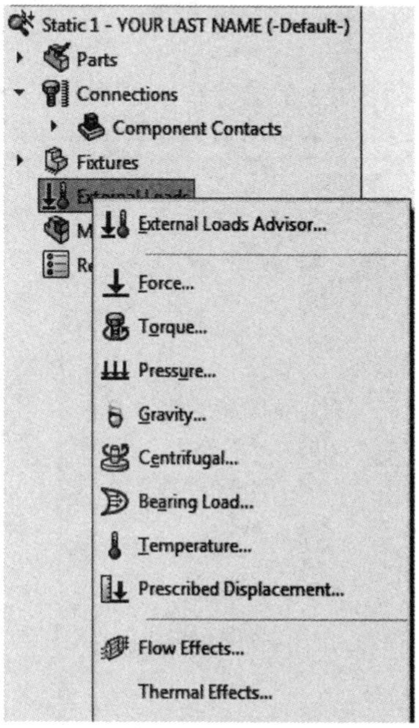

Figure 7-20. Applying External Loads.

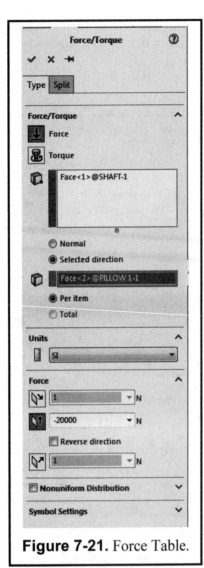

Figure 7-21. Force Table.

4. For the Second Selection box, pick the **Front Vertical surface of the Pillow Block**.
5. Set the units to the **SI** system.
6. Under **Force,** activate the second force and enter the value of the force **(-20000 N)** and click **OK (Figure 7-21)**. Forces appear as lavender arrows. *Note: Make sure that the lavender arrows are pointing down.*

This completes the application of restraints, and forces (**Figure 7-22**) to the Pillow Block and the Shaft. The next step is to create the Mesh required for analysis.

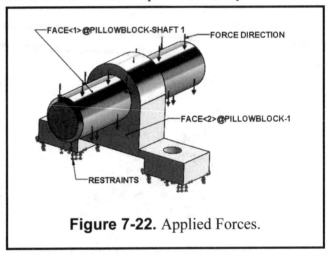

Figure 7-22. Applied Forces.

Creating the Mesh:

The mesh is the division of the solid into a number of small geometric divisions. The effect of the stress on these divisions is computed and then integrated to provide a stress analysis of the entire part. This is an extremely simplistic way of looking at the mesh generation process, which in reality is extremely complicated and takes a relatively long time to execute even on today's fastest computers. The procedure to create the mesh for this analysis is detailed below:

In the SOLIDWORKS Simulation Manager tree, click the **RMB** on **Mesh** and click on **Create Mesh (Figure 7-23)**.

1. This opens up the mesh dialog box, which allows the user to define the mesh size for the analysis. A smaller mesh yields better results but takes longer to mesh as well as analyze. Hence it is a tradeoff between mesh size and computation time. Change the units to **mm**. Set the **Global Size** to **5** for a slightly finer mesh and **Tolerance** to **.25 (See Figure 7-24)**.

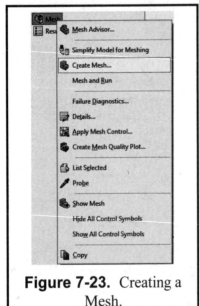

Figure 7-23. Creating a Mesh.

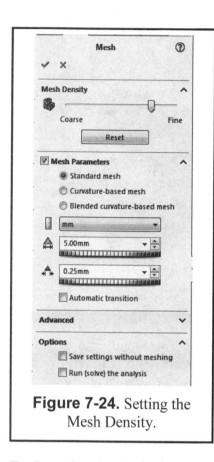

Figure 7-24. Setting the Mesh Density.

2. Click on **OK** to accept the values and the Meshing process begins.

3. At the end of the meshing process, a check mark is placed against Mesh in the SOLIDWORKS Simulation Feature Manager Tree. This causes the mesh to be displayed on the solid as shown in **Figure 7-25**.

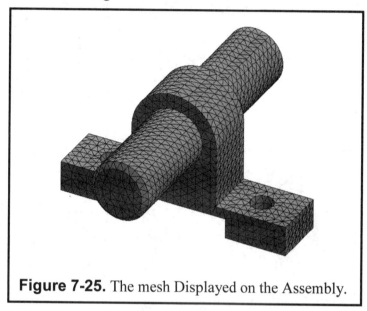

Figure 7-25. The mesh Displayed on the Assembly.

To Run Static Analysis

The mesh generation is now complete and all parameters required for running the analysis have been defined. To run the Static Analysis, click the **RMB** on the **Name of the Study** and click on **Run**. The analysis begins. This process also takes a long time to complete and the execution time depends on the mesh size. The program displays a message once the static analysis is completed. Click on **OK**.

Post-processing

In this section the analysis has been completed and the results are visualized and interpreted. If the study ran successfully, a new item (**RESULTS**) will appear in the Manager tree. Click on the + sign to expand this item. Three new items now appear in the **SOLIDWORKS Simulation** feature manager (**Figure 7-26**). They are **Stress, Displacement,** and **Strain**. You can right-click on any of these menu items and pick the action you wish to take (See **Figure 7-27**). **SHOW** illustrates what the study has computed, **DELETE** is obvious, and **COPY** puts the results on the computer clipboard. **Right Mouse Click** on **STRESS1** and select **Show**. You should get a figure that resembles **Figure 7-28**.

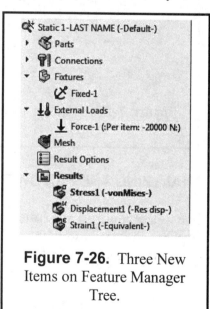

Figure 7-26. Three New Items on Feature Manager Tree.

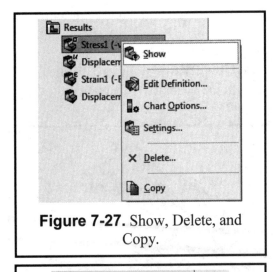

Figure 7-27. Show, Delete, and Copy.

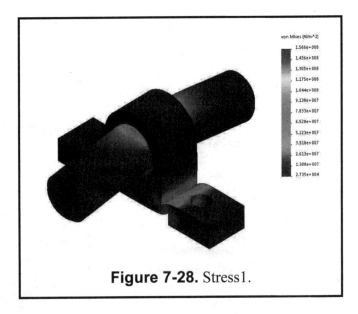

Figure 7-28. Stress1.

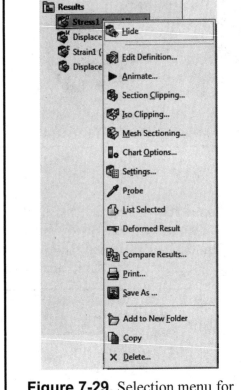

Figure 7-29. Selection menu for Stress1.

Right-click on **STRESS1** a second time to get a selection window as shown in **Figure 7-29**. Select several of the options to see what results are generated. If you select the **Edit Definition** you will notice that the Scale Factor is approximately 357. The distortion of the image on the computer is exaggerated by a factor of 357. If you select **Animate** the image will be put into motion. **Section Clipping** lets you manipulate a cross section to see internal forces and distortions.

Go to the **Stress1** Selection menu and choose **Print.** Go to **Properties** and change the print format to **Landscape**. Then click on **OK** and the plot will be printed out.

Insert the **Pillow-1 Assembly** rendered image onto a **Title Block** drawing sheet that was created in Chapter 1 and **Print** it on this sheet. This sheet along with the print of the **Stress1 - Von Mises** are to be turned in to your instructor.

ANALYZING THE PILLOW BLOCK FOR STRESSES

One of the powerful applications of solid modeling using SOLIDWORKS is the ability to easily interface the database with analysis software. In this case, the Pillow Block was analyzed for stress distributions when forces were applied to it during its normal function, which is to support a rotating shaft.

The FEA study determines how the stresses and strains were distributed throughout the solid material of the Pillow Block when the given forces were applied. The results are readily displayed as color contour maps that show the stress distribution. Areas where stresses concentrate can be detected and the FEA results can lead to an improved design of the part. **Figure 7-28** shows the results of this FEA study. It can be seen that the stresses tend to concentrate in the thin cylindrical wall surrounding the shaft hole, and also at the right angle where the upright feature emerges from the Pillow Block base. These two design features are likely candidates for further consideration and re-design when you create the next version of the Pillow Block. With the results of all this analysis, you are now ready to start the design modification stage.

DESIGN MODIFICATION OF THE PILLOW BLOCK

After studying the results of the FEA, several modifications are suggested for the pillow block to create a second version.

EDITING THE PILLOW BLOCK

Make sure the former **Pillow1** part file is **OPEN** on your computer screen, and then **SAVE AS Pillow2** to start a new version of the part. You will now see how easy it is to make most of the design modifications by simply editing the sketch that was used to extrude the base part. In the Feature Manager tree, click the **+** sign on the **Base/Extrude** step to expose the **Sketch1** icon on the tree. Move the mouse cursor over that **Sketch1** icon and click it with the right mouse button **(RMB)**. This invokes a pop-up menu that allows you to edit that sketch. On the pop-up menu, select the **Edit Sketch** option and click it with the **LMB**. Using either **Table 2** "Design Modification Table – A" and/or **Figure 7-30**, make the changes to modify the sketch to conform to the new criteria. When this is completed repeat the above process on the **Cut-Extrude** to relocate the holes using **Table 2 "Design Modification Table – A"** and/or **Figure 7-31**.

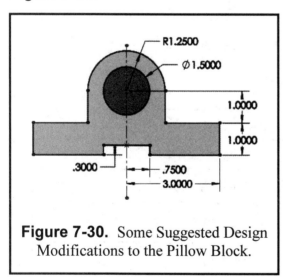

Figure 7-30. Some Suggested Design Modifications to the Pillow Block.

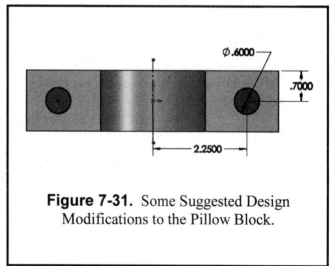

Figure 7-31. Some Suggested Design Modifications to the Pillow Block.

Table 2. Design Modification Table – A.	
Feature	**Change Order Request**
Outer Radius Around Shaft Hole	Change Radius from 1.125 to 1.25 inches
Change the thickness of the Base	Change Height from .75 inches to 1.00 inch
Center of the Shaft Hole Distance from Base	Change Height from 1.50 to 1.00 inch
Vertical Bolt Holes	Change Position from 2.00 to 2.25 inches from the centerline

Basically, the thickness of the material around the shaft hole was too thin and the outer radius of the pillow block needs to be increased in order to thicken the material surrounding the hole. Sharp 90 degree angles are generally not a good idea for mechanical design, and fillets can be added to bolster the strength at these junctures. Also the slot, which is used primarily to save on material waste and part weight, is poorly designed. The width should be reduced and fillets added to compensate for the downward pressure. Finally the vertical bolt holes (**Figure 7-31**) need to be centered on the new foot-pad bases.

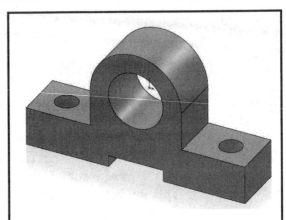

Figure 7-32. Some Modifications to the Pillow Block.

Once all the current changes have been made, select **Rebuild**. The new Pillow2 will appear in its current state (**Figure 7-32**).

After the design has been updated to look like **Figure 7-32**, **add fillets** to the Pillow block according to **Table 3** –"**Design Modification Table - B.**" When this is completed, your object should look like the Pillow Block in **Figure 7-33**.

Table 3. Design Modification Table - B	
Feature	**Change Order Request**
Right Angle Surface Intersections (2 Places)	Add Fillets with .375 inch Radius
Bottom Slot – Upper Two Corners	Add .25 inch Fillets
The Upper Edges of the Base	Add .25 inch Fillets

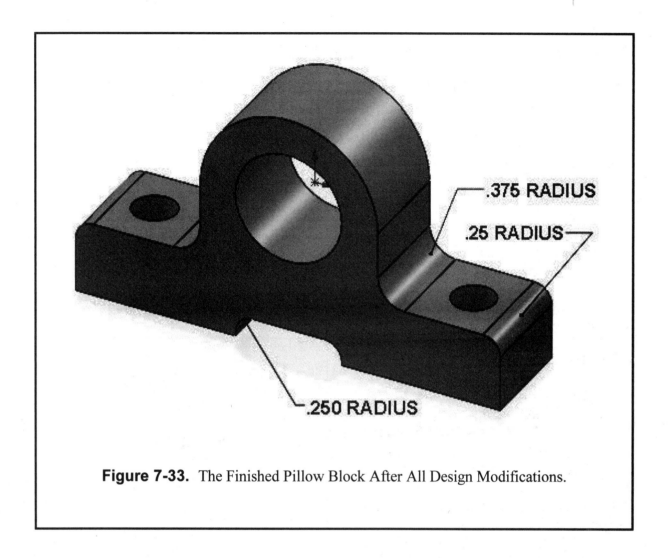

Figure 7-33. The Finished Pillow Block After All Design Modifications.

Study – 2 – LAST NAME

After you have finished the design modifications, go to **File**, select **New.** Click on **Assembly** and **Save As "Pillow Assembly-2."** Repeat the assembly process you used to create the first assembly. The two parts that make up the new assembly are **Pillow-2** and the **Shaft**.

You will now **Repeat** the **SOLIDWORKS Simulation** process, using the same criteria and procedure performed on the original design. **Save** the **Assembly** and the SOLIDWORKS **Simulation/Study2 – Your last name** to your folder. Insert the **Pillow-2 Assembly** rendered image onto a **Title Block** drawing sheet that was created in Chapter 1 and **Print** it on this sheet. This sheet along with the prints of the **Stress1 - Von Mises** for the second study are to be turned in to your instructor.

Exercise 7.2: Thermal Analysis of a Computer Chip

A computer chip is one of the many components in a computer, yet it contributes significantly to the very thing that may damage its performance. Computer designers must consider temperature, fatigue, and chemical and material properties among the many factors that may affect the performance. Heat has to be a major consideration when designing a computer system. As heat builds up in a component, the function and integrity of that component and even the entire system may be affected.

In this exercise you will construct a 3D solid model of a computer chip and a heat sink to see how one affects the other. Then the material properties, loads, reactions, and constraints typical for such a part will be presented. Finite element analysis will be used to visualize the resulting distributed heat through the chip and then with the heat sink attached. The first portion of this exercise will be to do a steady state study. You will then use the results to do a Transient thermal analysis.

CONSTRUCTING THE HEAT SINK

We begin with a sketch for the Heat Sink. Open **ANSI-INCHES.prtdot** from your folder. Do a **Save As – Heat Sink.sldprt**. Under the **Tools – Options – Document Properties** select **Units** and change them to **7 decimal places**. Select **Front** as the sketch plane, go to **Sketch** mode and draw a **Vertical Centerline** to the left of the origin. Go to **TOOLS - SKETCH TOOLS** and **SELECT – DYNAMIC MIRROR.** Now **SELECT** the **LINE** command and draw a angled line downward and another very short horizontal line to the centerline. Dimension the lines according to the sketch in **Figure 7-34** Add a Relation so that the end points of the two angled lines are horizontal with the Origin. A detail of the very small end of the pattern is shown in **Figure 7-35**. When finished, select the **Linear Sketch Pattern**. Set **Distance-1** to **.3125** and the **Instance Count** to **7**. In the box labeled **Entities to Pattern** at the bottom of the Linear Pattern display, **Select** the two **Diagonal Lines** and the very **Short Bottom line** to be arrayed. The Array should extend over the Origin. The completed model will have the Origin on the top surface and in the center of the model. This will help in the assembly of the parts for analysis. The next step is to

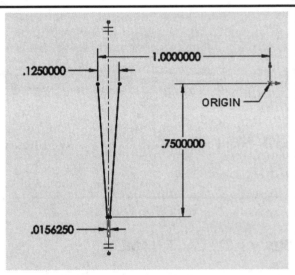

Figure 7-34. Base Drawing of the Fin for the Heat Sink.

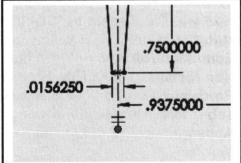

Figure 7-35. Detail of the Small End of the Fin Profile.

draw a **Horizontal Line** all the way across the top of the pattern. Then **Offset** that horizontal line by **.125** inches downward. You will have to un-select "Select Chain" for this to complete. Your resulting figure should look like **Figure 7-36**.

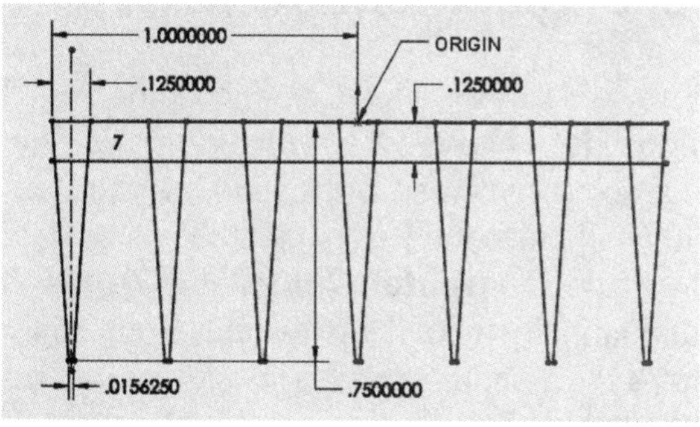

Figure 7-36. The Two Horizontal Lines.

Next, **Trim** the **angled lines** above the **offset line** and the **offset lines** between the **angled lines**. Be sure to also **Trim** the two very short segments of the offset line that extend beyond the outer edges of the pattern. Your completed sketch should appear as in **Figure 7-37**.

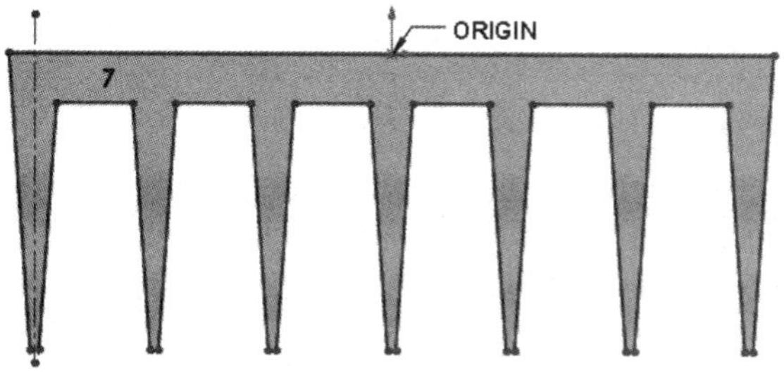

Figure 7-37. The Trimmed Pattern.

To complete the Heat Sink, **Extrude** the Sketch – **Mid-Plane** for **2** inches. Your solid model should look like the model in **Figure 7-38**. You will now assign the material to the Heat Sink. **Right click** on **Material <not specified>** in the Feature Manager Tree, **Edit Material**, expand **Aluminum Alloys** and select **1060 Alloy**. Also activate the **Appearance Tab** to make sure that the "**Apply Appearance of: 1060 Alloy**" is **unchecked**. Click **Apply** toward the bottom of the window and then **Close** your Material Window. You have now completed the model of the Heat Sink. You may assign a color to the Heat Sink. Save your model in your File Folder as **Heat Sink.sldprt**.

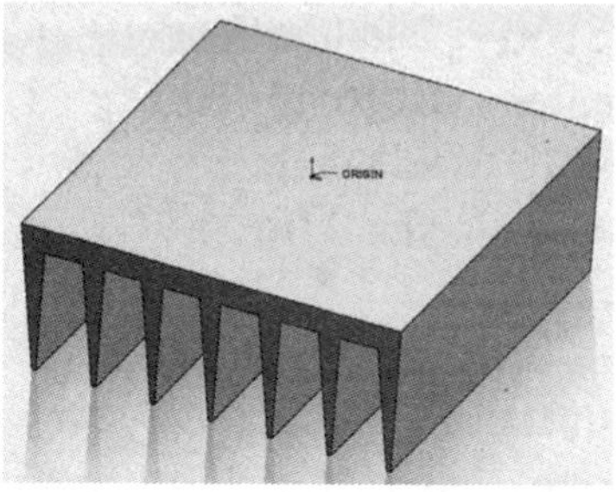

Figure 7-38. The Finished Heat Sink.

CONSTRUCTING THE CERAMIC COMPUTER CHIP

Begin with a sketch for the Ceramic Computer Chip. Open **ANSI-INCHES.prtdot** from your folder. Do a **Save As CERAMIC COMPUTER CHIP.sldprt**. Select **Top** as the sketch plane, go to **Sketch** mode, and create the profile as shown in **Figure 7- 39**. This is a simple square that measures **1.50** inches centered at the **Origin**. Click on the Features Tab and **Extrude** the Sketch **Upward**

.125 Inches. **Right click** on **Material <not specified>** and **Edit Material**. Expand **Other Non-Metals** and Select **Ceramic Porcelain**. Under the **Appearance Tab** make sure the **Apply Appearance** box is unchecked. You may assign a color to the Computer Chip. Save your model in your File Folder as **Computer Chip.sldprt**.

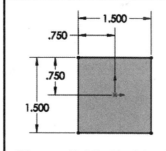

Figure 7-39. Sketch of Computer Chip.

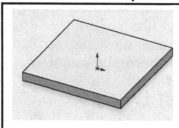

Figure 7-40. Completed Ceramic Computer Chip.

CONSTRUCTING THE ASSEMBLY

With the Heat sink and Ceramic Computer Chip solids open, go to **File – New – Assembly**. In the Feature Manager Tree select the **Heat Sink** and **Okay** to Begin Assembly. This will place the Heat sink at the Origin. Next, **Insert** the **Ceramic Computer Chip** at the **Origin**. This will automatically mate the two solids to finish the assembly. **Save** the assembly to your file folder as **Thermal Study.sldasm**. Your assembly should look like **Figure 7-41**.

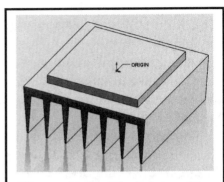

Figure 7-41. Final Assembly.

THERMAL ANALYSIS OF THE CERAMIC COMPUTER CHIP

Starting the Study

Go to the **Tools** tab – **Add-Ins** and select **SOLIDWORKS Simulation** and click **OK**. This will insert a pull down tab entitled **Simulation**. Activate the Simulation ribbon menu. Under **Study Advisor** select **New Study**. In the box labeled **Name**, enter **Thermal Study 1 - (your last name)**. In the **Type** box activate the **Thermal** Icon as shown in **Figure 7-42**. **Okay** your settings.

In the Thermal **Study** manager tree, right click on **Connections** and Select **Contact Set**. Since you only have two pieces and they are in contact with each other, you can **Click** on "**Automatically find Contact Sets**." Under Options **Select Touching Faces**. Under Components **Select** the **Two Parts** of the Assembly. Under Results make sure that the **Type** is set to "**Thermal Resistance**." Under Thermal Resistance **Select - Distributed** and insert the **SI** units of **.0001 K-m^2/W**. Under the **Advanced** box select **Surface to Surface**. See **Figure 7-43** for the settings. **Okay** all of the settings.

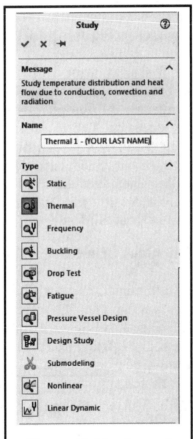

Figure 7-42. Thermal Study Set-up.

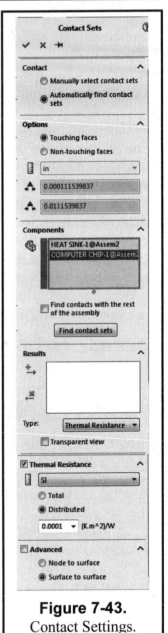

Figure 7-43. Contact Settings.

Identifying Heat Source and Convection Settings

The heat produced in a computer chip will vary depending on its complexity. For this exercise, 15 Watts of power will be used. The next phase of this study is to identify the amount of heat and what is creating the heat. Right click on **Thermal Loads** and select **Heat Power**. This opens a panel in the Feature Manager area for input. **Right Click** on the **Heat Sink** and **Hide** it so you can rotate the **Computer Chip** to **Select** the **Bottom** surface. This is the surface that will be transferring the heat to the Heat Sink. Your Surface selection will appear in the light blue box. Under the **Heat Power** enter **15 Watts**. When completed **Okay** the selections. See **Figure 7-44** for the settings. Before you proceed to the next step, **Right Click** on the **Heat Sink** and **Click** on "**Show.**"

The last step before the mesh is created is to identify the surfaces that will disburse the heat from the Computer Chip. **Right Click** on **Thermal Loads** and select **Convection**. Now **Select** all of the surfaces of the **Heat Sink except** for the **Top Surface** (**there are 29 Surfaces that should be selected**). Now set the **Convection Coefficient** to **15 W/(m^2.K)** and set the **Bulk Ambient Temperature** to **300** degrees Kelvin. These settings can be seen in **Figure 7-45**.

Creating the Mesh

The Mesh is the method of dividing the solid model into an ordered set of finite geometric elements that can be mathematically analyzed to determine the effects of stress, strain or temperature on a solid.

Right Click on **Mesh** in the Feature Manager Tree and select **Create Mesh**. Under the **Mesh Parameters** change the **Units** to **mm**, the **Global size** to **2.00mm** and the **Tolerance** to **0.10mm** as shown in **Figure 7-46**. When you select the check mark to okay the settings, the program will Mesh the assembly as shown in **Figure 7-47**.

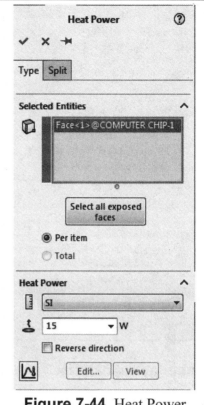

Figure 7-44. Heat Power Settings.

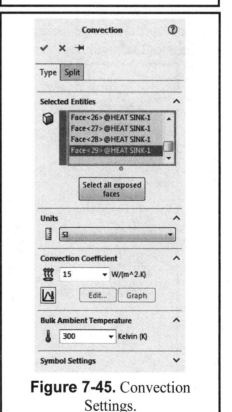

Figure 7-45. Convection Settings.

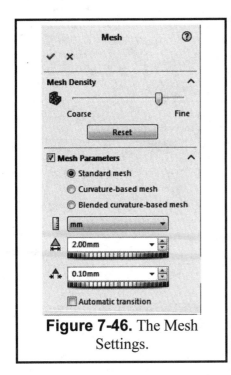

Figure 7-46. The Mesh Settings.

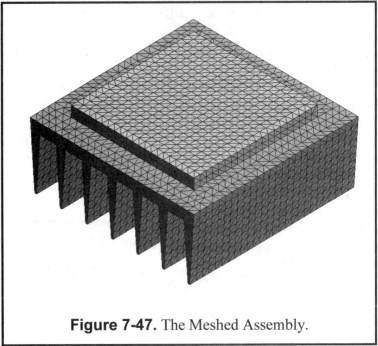

Figure 7-47. The Meshed Assembly.

Running the Analysis

Right Click on the name of the study in the **SOLIDWORKS Simulation** Feature Manager and select **Run**. There will be some delay (up to a Minute) in getting the results of the study because of the amount of calculations needed to complete the study. The Feature Manager Tree will now display a new listing identified as **Results: Thermal1 (-Temperature-)** along with an image of the results of the study (See **Figure 7-48**).

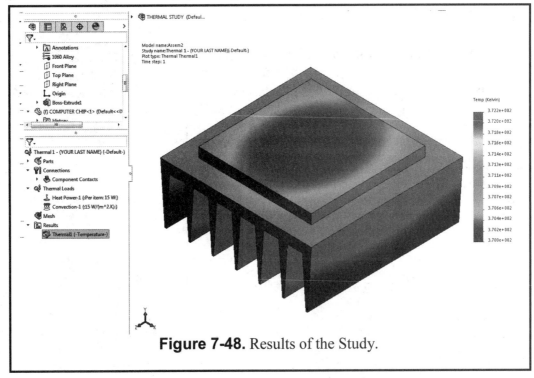

Figure 7-48. Results of the Study.

Right click on **Thermal1 (-Temperature-)** and select **Print** to get a hard copy of your Thermal Study.

The next step is to **Right Click** on **Thermal1 (-Temperature-)** and select **Edit Definition**. Under **Display** activate the Down Arrow of the **TEMP: Temperature** box and **Select** the last option – **HFLUXN: Resultant Heat Flux, Figure 7-49**. Then under **Advanced Options**, check the box next to **Show as vector plot**.

It is possible to adjust the density and size of the Vector Plot. **Right Click** on **Thermal1 (-Res heat flux-)** and **Select - Vector Plot Options**. Enter **500** in the first box and **25** in the second box (**See Figure 7-50**). You will then get a result that shows the direction of heat flow (**Figure 7-51**). There are arrows going straight down out of the Computer Chip and disbursing out through the fins of the Heat Sink. **Right Click** on **Thermal1** and **Print** a copy of the Heat Disbursement Illustration.

Insert your **Assembly** on a **Title sheet** with the usual annotations and submit this along with the prints of **Thermal1 (-Temperature-)** and **Thermal1 Vector Plot** to your instructor.

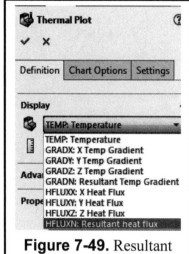

Figure 7-49. Resultant Heat Flux.

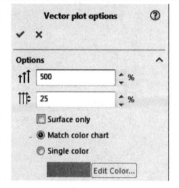

Figure 7-50. Vector Plot Options.

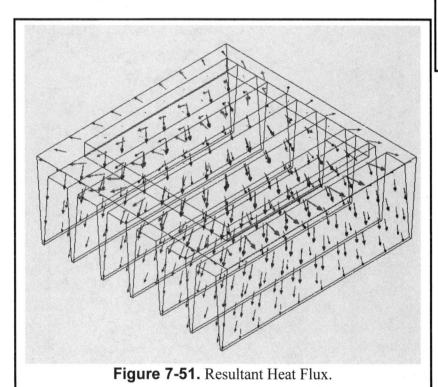

Figure 7-51. Resultant Heat Flux.

NOTES:

Computer Graphics Lab 8:
Kinematics Animation and Rapid Prototyping

In Computer Graphics Lab 8, you will explode a previously built assembly file using SOLIDWORKS "Assembly Exploder." You will set the direction and distance of the exploded parts of the assembly. You will then create a simple animation of this exploding assembly using SOLIDWORKS "Animation Wizard." You will save the animation as an .AVI file that can be played on an external viewer, like Windows Media Players. You will also be briefly introduced to SOLIDWORKS Physical Simulation capabilities. A second objective of this Computer Graphics Lab 8 is to produce an .STL (stereo lithography) file of your SOLIDWORKS parts. The .STL files are then transferred to a rapid prototyping machine to make physical prototypes of the parts.

ASSEMBLY TOOLBAR

In the earlier Computer Lab 5, you were introduced to the **Assembly Toolbar** as shown in **Figure 8-1**. When you open a completed assembly file, you can use these tools to manipulate the assembly components. When you click on the **Exploded View** icon, the **Explode** menu appears on the screen, as shown in **Figure 8-2**. This menu can be used to explode the assembly by manually editing the explode process.

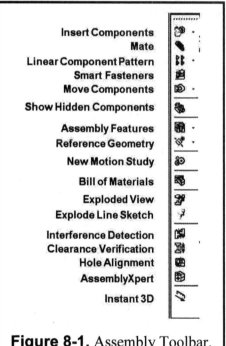

Figure 8-1. Assembly Toolbar.

Figure 8-2. Exploded Menu.

For example, in **Figure 8-3** the **Pin** of the Terminal Support assembly has been exploded upward as shown. The user first would highlight the **Settings** window tab (it is light blue) and then pick the **Pin** part on the assembly. The Pin part identification (like *pin-1@ wing base*) now appears in the window. Three direction arrows appear on the part picked. Highlight the direction arrow (like *Y@ wing base*) that you want the part to move. Then key in the **Distance** to be **6.00** inches. Then toggle the **Direction** button so the Pin goes upward. To preview, click the **Apply** button. To finish, click the **Done** button and *Explode Step 1* is done.

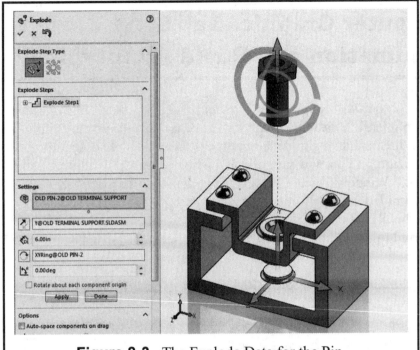

Figure 8-3. The Explode Data for the Pin.

ANIMATION WIZARD

The "Animation Wizard" is an easy tool to create simple animations when you have a part or assembly file loaded. Activate the Assembly menu and select **New Motion Study** (See **Figure 8-4**).

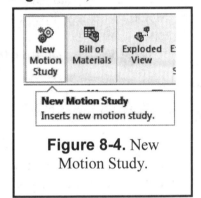

Figure 8-4. New Motion Study.

An animation Controller window appears in the lower part of the screen. Select the Animation Wizard icon (See **Figure 8-5**). It has three features (See **Figure 8-6**):

      Rotate Model
      Explode Model
      Collapse Model

If you select **Rotate Model**, it asks for an axis to rotate around and the number of rotations. You then get to the next menu (see **Figure 8-7**), which asks about the duration (in seconds) for the animation. In order to select the other two choices in the "Animation Wizard" you must first **Insert** an **Exploded View** of an assembly. You can then create an **Explode** animation or a **Collapse** (reverse of explode) animation.

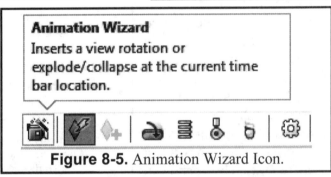

Figure 8-5. Animation Wizard Icon.

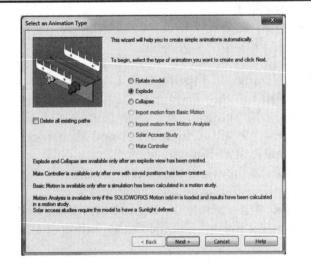

Figure 8-6. The Animation Wizard Create Menu.

ANIMATION CONTROLLER AND SCRIPT

The "Animation Controller and Script" appears on the bottom of the screen **(Figure 8-8)** when you select a **New Motion Study** as illustrated in **Figure 8-4**.

Figure 8-7. The Animation Control Options Menu.

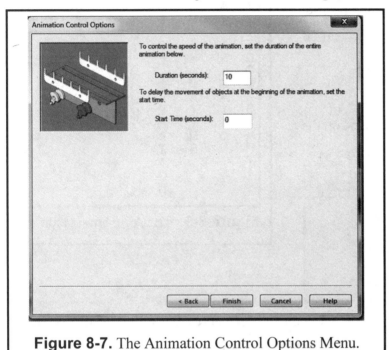

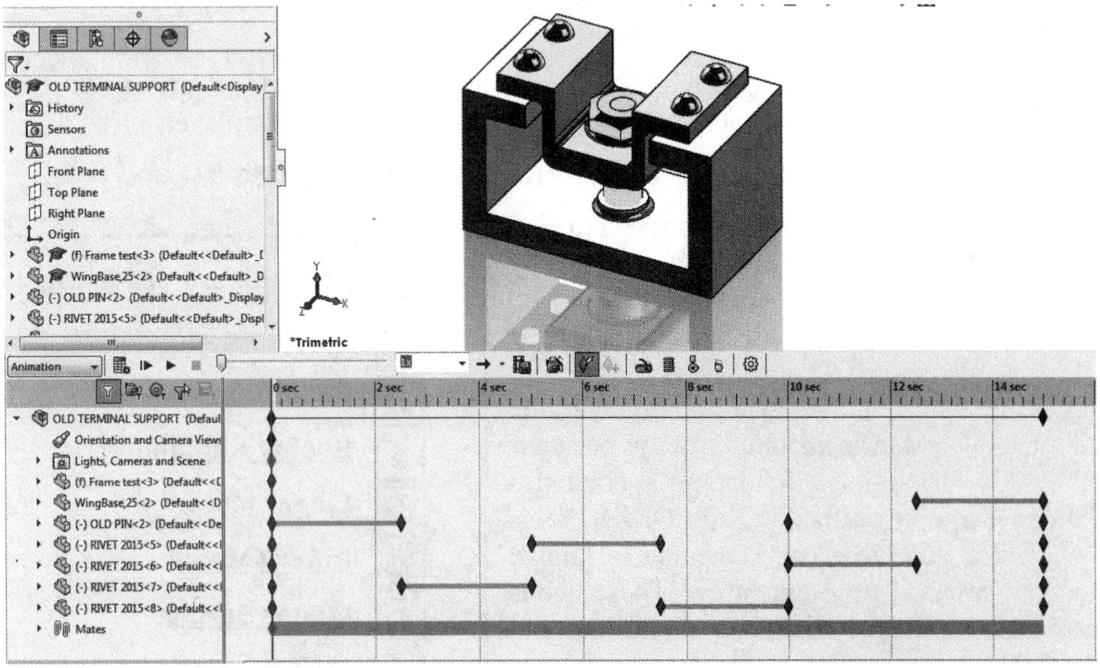

Figure 8-8. The Animation Controller and Script Area.

The following button commands are available on the "Animation Controller" (**Figure 8-9**).

Play from Start button stops the animation and sends it back to play from the beginning position.

Play button plays the animation at the specified speed.

Stop button stops the animation and sends it back to the beginning position.

There are three "**Modes of Operation**" on the controller.

Figure 8-9. The Animation Controller.

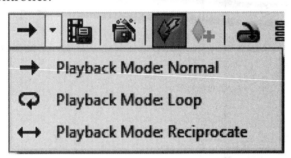

Normal Mode button sets the animation to run once from start to end at the specified speed.

Loop Mode button sets the animation to run over and over again in a continuous loop.

Reciprocate Mode button sets the animation to first run forward, and then run backwards, and repeat this reciprocation in a continuous fashion.

Save as AVI File button will open the "Save Animation to File" menu and allows you to save the animation as an .AVI file. It can then be played on an external media player.

Animation Wizard launches the "Animation Wizard" tool (see **Figures 8-4 and 8-5**).

Zoom In and **Zoom Out** buttons allow you to zoom the script in or out on the screen.

INTRODUCTION TO PHYSICAL SIMULATION

Physical Simulation allows you to simulate the effects of motors, springs, and gravity on your assemblies. Physical Simulation combines simulation elements with SOLIDWORKS tools such as mates and Physical Dynamics to move components around your assembly. These icons appear on the top of the Animation Controller and Script area.

The icons on the Simulation Toolbar are shown in **Figure 8-10**. These options include the ability to add a Linear Motor, Rotary Motor, a Linear Spring, or Gravity to your model.

Figure 8-10. The Simulation Toolbar.

Exercise 8.1: Exploded Animation of the TERMINAL SUPPORT ASSEMBLY

Recall in Computer Graphics Lab 5 you created some assembly models in SOLIDWORKS. If you created and saved the Terminal Support Assembly, you can now work this Exercise 8.1. If you created and saved the Pulley Assembly, you should go to Exercise 8.2. If you did not save either assembly file, then you need to return to Computer Graphics Lab 5 and re-build one of the assemblies first and then return to this Lab 8.

EXPLODING THE ASSEMBLY

Pull down **File** and **Open** the previous **Terminal Support.sldasm** assembly file in your designated folder. You will create the animation by first exploding the assembly. To do this, choose the **Insert** menu and click on **Exploded View** (or click the Exploded View icon on the Assembly Toolbar). The "Explode" menu now appears in the Feature Manager area as shown in the earlier **Figure 8-2.**

Now the first part to explode is the **Pin**. Click in the **Part** area of **Settings** window (it is light blue). Next, click on the **Pin** part of the model and the part gets selected using some SOLIDWORKS code like *Pin-1@terminal support assembly*. Go to the model and pick the **upward direction arrow** to get the correct direction, named by a SOLIDWORKS code like *Y@terminal support assembly*. Then key in the **Distance** to be **8.00** inches. Then toggle the **Direction** button so the Pin goes upward. To preview, click the **Apply** button. To finish, click the **Done** button and *Explode Step 1* is done. These correct explode settings for the Pin part are shown in **Figure 8-11**.

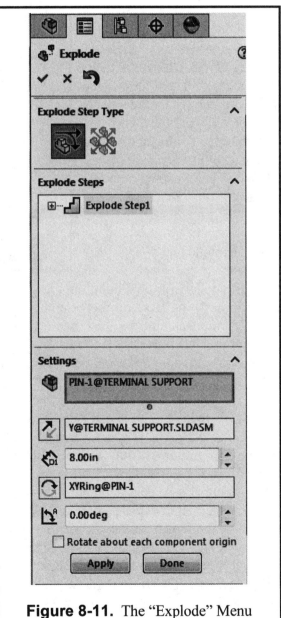

Figure 8-11. The "Explode" Menu Settings for the Pin.

Now explode the four Rivets either individually or together. Click in the **Part** area of **Settings** window (it turns light blue). Pick the four Rivets. They appear in the Part window using some SOLIDWORKS code like *Rivet-1@terminal support assembly*, etc. Go to the model and pick the **upward direction arrow** to get the correct direction, named by a SOLIDWORKS code like *Y@terminal support assembly*. Then key in the **Distance** to be **7.00** inches. Then toggle the **Direction** button so the Rivets go upward. To preview, click the **Apply** button. To finish, click the **Done** button and *Explode Step 2* is done. These correct explode settings for the Rivet parts are shown in **Figure 8-12**.

Finally, explode the Wing Base part. Click in the **Part** area of **Settings** window (it turns light blue), and then pick the Wing Base on the assembly model. It appears in the Part window using some SOLIDWORKS code like *Wing Base-1@terminal support assembly*. Go to the model and pick the **upward direction arrow** to get the correct direction, named by a SOLIDWORKS code like *Y@terminal support assembly*. Then key in the **Distance** to be **3.00** inches. Then toggle the **Direction** button so the Wing Base goes upward. To preview, click the **Apply** button. To finish, click the **Done** button and *Explode Step 3* is done. These correct explode settings for the Wing Base part are shown in **Figure 8-13**. The Terminal Support is now exploded, as shown in **Figure 8-14**.

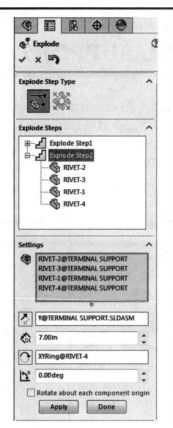

Figure 8-12. "Explode" Settings for the Rivets.

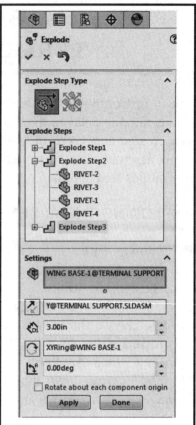

Figure 8-13. "Explode" Settings for the Wing.

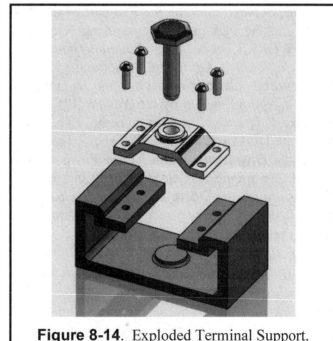

Figure 8-14. Exploded Terminal Support.

CREATING THE ANIMATION

Now you will create the animation using the automatic Animator Wizard. Click on **NEW MOTION STUDY** under the Assembly tab in the top menu bar. The "Animator Controller and Script" area appears on the bottom of the screen as shown in earlier **Figure 8-8**. Select the

Animation Wizard  button on the Controller (see the choices on **Figure 8-6**). Next, click **on** the **Explode** on the "Select an Animation Type" menu. Then click the **Next** button. In the next dialog menu (see **Figure 8-7**) enter the animation "Duration" of **15 or more** seconds and "Start Time" as **0**. Finally, click the **Finish** button.

PLAYING THE ANIMATION

Study the various buttons on the "Animation Controller" as shown in earlier **Figure 8-8**. Press the **Play** button and see the animation on your screen. Try some of the other animation buttons. **Play** the animation in the **Normal Mode**. **Play** the animation in the **Loop Mode** and see the assembly move in a repeated fashion until you click **Stop**. **Play** the animation in the **Reciprocate Mode** and see the assembly move in a reciprocating mode until you click **Stop**. Try several combinations of options for this exploded animation (see **Figure 8-15**). When you have selected **Normal Mode**, show it to your instructor or lab teaching assistant.

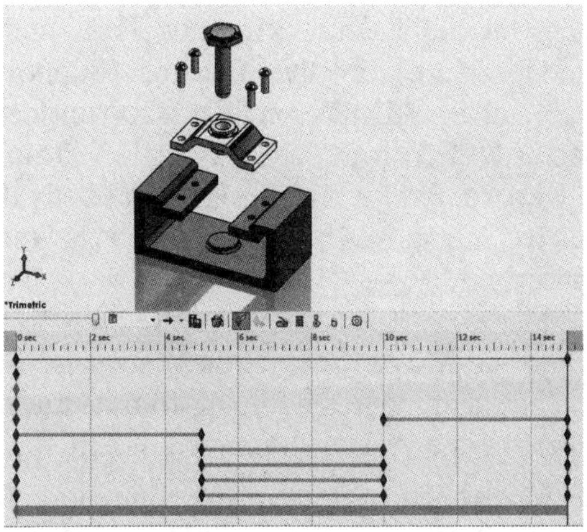

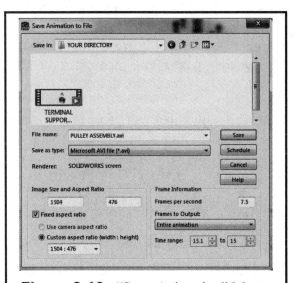

Figure 8-15. The Terminal Support Animation Script.

SAVING THE ANIMATION

Return to the "Animation Controller" buttons and select the **Save** button (). The "Save Animation to File" menu now appears on the screen as shown in **Figure 8-16**. Name the file **Terminal Support** and use an **.AVI file** type. Use the **SOLIDWORKS screen** as the Renderer and set the frames per second rate to **15**. Then click the **Save** button. Click on **OK** in the "Video Compression" dialog box that next appears, in order to accept the default settings. This causes the exploded model animation to be recorded to your file as well as played on screen. It may move slowly at this point because it is saving data to a file. Note: You may have to adjust your model viewport to fit the whole animation on your screen as it is being captured.

Figure 8-16. "Save Animation" Menu.

At this point you are finished with the animation, so you may want to **Save** your exploded assembly file as **EXPLODED TERMINAL SUPPORT**.

PHYSICAL SIMULATION EXAMPLE

Return to the original Terminal Support file. Pull down **File** and **Open** the previous **Terminal Support.sldasm** assembly file in your designated folder. Make sure it is collapsed and in the original mating position. If it is not collapsed, select the **Move Component** icon on the Assembly Toolbar, and move the Pin down to its mated position. Now you need to remove one of the mating constraints. Click the **Mates** tab ⊞ in the Feature Manager Tree, and see all the mated features for the Terminal Support Assembly. Right click (RMB) on the **Coincident(X) Wing Base<1>,Pin** mate, and then select **Delete** as shown in **Figure 8-17**. Then answer **Yes** to the "Confirm Delete" menu. This allows you to move the Pin upward.

Figure 8-17. Deleting the Coincident Mate between the Pin and the Wing Base.

Select the **Move Component** icon on the Assembly Toolbar, and move the Pin up from its mated position to be above the rest of the mated assembly. Next select the **Motor** icon 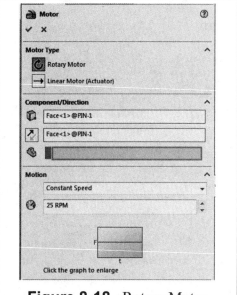 from the Simulation Toolbar (see **Figure 8-10**). Click on the cylindrical shaft of the Pin. Next, set the **RPM** below the Constant Speed to **25 RPM** as shown in **Figure 8-18**. Then click the green (√) check. Now select the **Calculate Simulation** icon button on the Simulation Toolbar. The computer will now calculate the physical simulation and play it on the screen as shown in **Figure 8-19**. **If instructed**, show it to your instructor and then press the **Stop Record** button on the Simulation Toolbar.

You can now click the **Replay Simulation** icon button on the Simulation Toolbar if you want. Now try to add a **Linear Motor** to one of the Rivets. You may have to first **Delete** a **Coincident** mate to get the Rivet to move away from the Wing Base. Then click the **Calculate Simulation** icon to start the new simulation.

Before you leave the lab, your instructor needs a hard copy of your work. So **Open** your previously saved assembly file, **Explode** it, and **Print** a color image of

Figure 8-18. Rotary Motor Menu.

the exploded assembly on a Title Block drawing sheet, as shown in **Figure 8-20**.

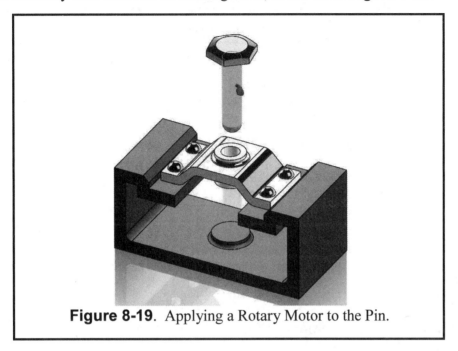

Figure 8-19. Applying a Rotary Motor to the Pin.

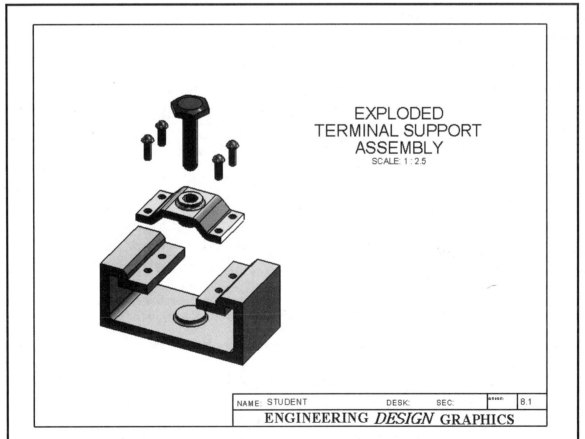

EXPLODED
TERMINAL SUPPORT
ASSEMBLY
SCALE: 1 : 2.5

NAME: STUDENT   DESK: SEC: DRSOF: 8.1

ENGINEERING *DESIGN* GRAPHICS

Figure 8-20. The Exploded Terminal Support Assembly Image on a Title Block Sheet.

Exercise 8.2: Exploded Animation of the PULLEY ASSEMBLY

Recall in Computer Graphics Lab 5 you created some assembly models in SOLIDWORKS. If you created and saved the Pulley Assembly, then you should work this Exercise 8.2.

EXPLODING THE ASSEMBLY

Read through the Kinematic Animation portion of Unit 8 (Pages 8-1 through 8-4) before you begin this exercise.

Pull down **File** and **Open** the previous **Pulley Assembly.sldasm** assembly file in your designated folder. You will create the animation by first exploding the assembly. To do this, choose the **Insert** menu and click on **Exploded View** (or click the Exploded View icon on the Assembly Toolbar). The "Explode" menu now appears in the Feature Manager area as shown in the earlier **Figure 8-2 on page 8-1**.

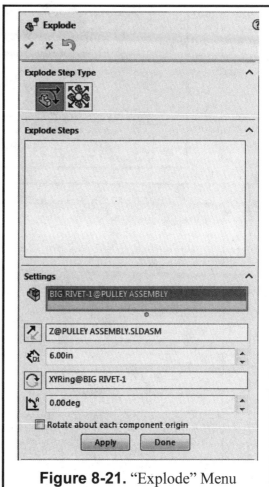

Figure 8-21. "Explode" Menu Settings for the Big Rivet.

Now the first part to explode is the Big Rivet. Click in the **Part** area of **Settings** window (it turns light blue). Next, click on the Big Rivet part of the model and the part gets selected using some SOLIDWORKS code like *Big Rivet-1@Pulley Assembly*. Go to the model and pick the **leftward direction arrow** to get the correct direction, named by a SOLIDWORKS code like *Z@Pulley Assembly*. Then key in the **Distance** to be **6.00** inches. Then toggle the **Direction** button so the Big Rivet goes leftward. To preview, click the **Apply** button. To finish, click the **Done** button and *Explode Step 1* is done. These correct explode settings for the Big Rivet part are shown in **Figure 8-21**. You should now see the Big Rivet move leftward 6 inches from the assembly.

In the same manner, you will now explode the pulley downward. Click in the **Part** area of **Settings** window (it turns light blue). Next click on the Pulley part of the model and the part gets selected using some SOLIDWORKS code like *Pulley-1@Pulley Assembly*. Go to the model and pick the **upward direction arrow** to get the correct direction, named by a SOLIDWORKS code like *Y@Pulley Assembly*. Then key in the **Distance** to be **6.00** inches. Then toggle the

Direction button so the Pulley goes downward. To preview, click the **Apply** button. To finish, click the **Done** button and *Explode Step 2* is done. These correct explode settings for the Pulley part are shown in **Figure 8-22**. You should now see the Pulley move downward 6 inches from the assembly.

You can complete the exploding of the remaining parts by repeating the above steps, one-by-one, for each part. These steps are as follows.

- Click in the **Part** area of **Settings** window (it turns light blue).
- Click on the desired **Part** in the assembly model.
- Click on the correct direction arrow.
- Key in the correct **Distance** value.
- Toggle the **Direction** button to get the correct explode direction.
- Click the **Apply** button to preview the explode.
- Click the **Done** button to complete the Chain.

Use the following data table for completing this Pulley Assembly explosion process. Note that the Spacer will stay in place and will not move in this explosion process.

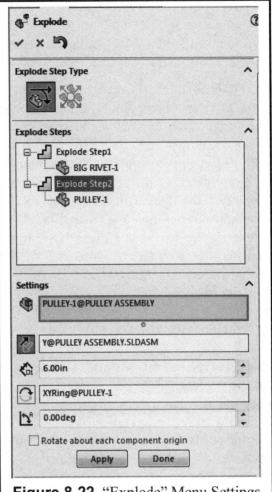

Figure 8-22. "Explode" Menu Settings for the Pulley.

Step	Component	Direction	Distance Value
Explode Step 3	Small Rivet 1	Leftward	6.00 inches
Explode Step 4	Small Rivet 2	Leftward	6.00 inches
Explode Step 5	Front Base Plate	Leftward	2.00 inches
Explode Step 6	Back base Plate	Rightward	3.00 inches
Explode Step 7	Eye Hook	Upward	3.50 inches

Once this table is completed, click the **OK** button to close the "Explode" menu. You should now have an exploded Swivel Eye Hook assembly as shown in **Figure 8-23**.

CREATING THE ANIMATION

Now you will create the animation using the automatic Animator Wizard. Click on **NEW MOTION STUDY**  under the Assembly tab in the top menu bar. The "Animator Controller and Script" area appears on the bottom of the screen as shown earlier in **Figure 8-5**, **page 8-2**. Select the **Animation Wizard** button on the Controller (see the choices on **Figure 8-5**). Next click **on** the **Explode** on the "Select an Animation Type" menu. Then click the **Next** button. In the next dialog menu (see **Figure 8-7**) enter the animation "Duration" of **15 or more** seconds and "Start Time" as **0**. Finally, click the **Finish** button.

Figure 8-23. The Exploded Pulley Assembly.

PLAYING THE ANIMATION

Study the various buttons on the "Animation Controller" as shown earlier in **Figure 8-8**. Press the **Play** button and see the animation on your screen. Try some of the other animation buttons. **Play** the animation in the **Normal Mode**. **Play** the animation in the **Loop Mode** and see the assembly move in a repeated fashion until you click **Stop**. **Play** the animation in the **Reciprocate Mode** and see the assembly move in a reciprocating mode until you click **Stop**. Try several combinations of options for this exploded animation. When you have selected the **Normal Mode**, show it to your instructor or lab teaching assistant. At this point, you may want to **Save** your exploded assembly file as **EXPLODED PULLEY ASSEMBLY.sldasm**.

SAVING THE ANIMATION

Return to the "Animation Controller" buttons and select the **Save** button. The "Save Animation to File" menu now appears on the screen as shown in **Figure 8-24**. Name the file **Pulley Assembly** and use an **.AVI file** type. Use the **SOLIDWORKS screen** as the Renderer and set the frames per second rate to **15**. Then click the **Save** button. Click on **OK** in the "Video Compression" dialog box that next appears, in order to accept the default settings. This causes the exploded model animation to be recorded to your file as well as played on screen. It may move slowly at this point because it is saving data to a file. Note: You may have to adjust your model viewport to fit the whole animation on your screen as it is being captured.

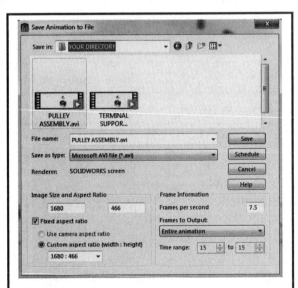

Figure 8-24. "Save Animation" Menu.

PHYSICAL SIMULATION EXAMPLE

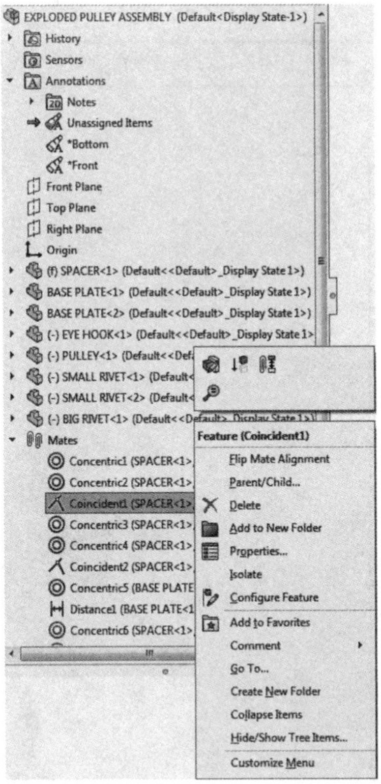

Return to the original **Pulley Assembly**. Pull down **File** and **Open** the previous **Pulley Assembly** file in your designated folder. Make sure it is collapsed and in the original mating position (see **Figure 5-46**). If it is not collapsed, select the **Move Component** icon on the Assembly Toolbar, and move the parts into their mated positions. Now you need to remove some of the mating constraints. Click the **Mates** tab ⊞ in the Feature Manager Tree, and see all the mated features for the Terminal Support Assembly. Right click (RMB) on the **Coincident(X) Spacer<1>, Base Plate<1>** mate, and then select **Delete** as shown in **Figure 8-25**. Then answer **Yes** to the "Confirm Delete" menu. Continue this process and **Delete** all the mates associated with the Front Base Plate, and all the mates associated with the three Rivets.

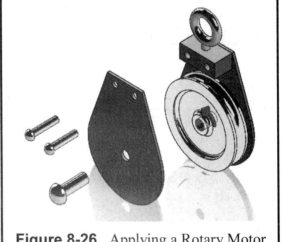

Figure 8-25. Deleting Mates.

Select the **Move Component** icon on the Assembly Toolbar, and move all three Rivets away from the assembly. Then move the front Base Plate away from the assembly. See **Figure 8-26** as a preview of the current screen arrangement. Next select the **Motor** icon 🖱 from the Simulation Toolbar (see **Figure 8-9**). Select the hole at the center of the Pulley. Next set the **RPM** below the Constant Speed to 20 RPM as shown in **Figure 8-18**. Then click the green (√) check. Now select the **Calculate Simulation** icon button on the Simulation Toolbar. The computer will now calculate the physical simulation and play it on the screen as shown in **Figure 8-26**. **If instructed** to do so, show it to your instructor and then press the **Stop Record or Playback** button on the Simulation Toolbar. You can now click the **Replay Simulation** icon button on the Simulation Toolbar if you want.

Now try to add a **Linear Motor** to one of the Rivets or the Eye Hook. You may have to first **Delete** a **Coincident** mate to get the Part to move away from the assembly model. Then click the **Calculate Simulation** icon to start the new simulation.

Figure 8-26. Applying a Rotary Motor to the Pulley.

Before you leave the lab, your instructor needs a hard copy of your work. So **Open** your previously saved assembly file, **Explode** it, and **Print** a color image of the exploded assembly on the Title Block drawing sheet, as shown in **Figure 8-27**. Save your drawing as **Pulley Assembly.slddrw**.

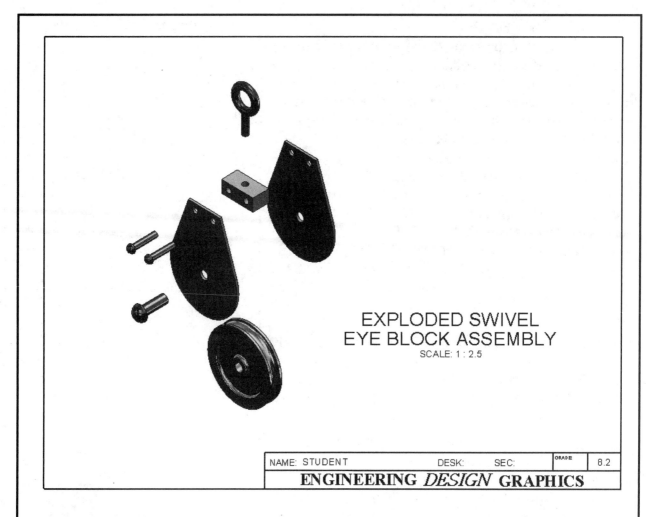

EXPLODED SWIVEL
EYE BLOCK ASSEMBLY
SCALE: 1 : 2.5

| NAME: STUDENT | DESK: | SEC: | GRADE | 8.2 |

ENGINEERING *DESIGN* GRAPHICS

Figure 8-27. The Exploded Pulley Assembly Image on a Title Block Sheet.

Exercise 8.3: Creating Views That Will Be Helpful When Illustrating Parts

In Exercise 8.3, you will be inserting part drawings onto a Titlesheet and arranging special views to show certain details more clearly. You will be activating the Insert Tab and using the Drawing View option.

To begin, open the **Toe Clamp.sldprt** file that you generated in **Exercise 3.4**. Also open your **Titleblock Inches.slddrw**. Activate the Window Tab and Tile Vertically. You will see both open files next to each other. Now go to the **Insert Tab – Drawing View – Model view**. Since the Toe Clamp is already open, it shows in the box titled **Part/Assembly to Insert** box. Click on the small **White circle with the Blue arrow** in it. In the Model View box in the Feature Manager area, check the **Create Multiple Views** box, then activate the **Top and Pictorial** views. Select the Green Check mark to place the views on the Titlesheet. If you activate the blue dashed box around the pictorial you will have a series of options to manipulate the image. Click on the second blue box from the right and you will get a shaded image. Just below the shaded box you will see Scale. Check the **Use Custom Scale** and change the scale to "**User Defined**" and enter **1:3**. Take your cursor just inside the blue dashed box and it will turn into a four directional arrow. With the left mouse button depressed you can move the image to the upper right corner of your drawing sheet. Also move the top view upward and to the left.

With the top view activated, click on the second box from the left in the Display Style to turn on the hidden lines. Now go to the **Insert – Annotations – Center Mark**. Go down to Slot Center Marks and pick the icon second from the left, click on the **largest arcs** of the slot and accept the operation. Then go to **Insert – Annotations – Centerlines. Check the Auto Insert – Select View** box. Now click on the Blue dashed box around the top view. This will complete the top view as shown in **Figure 8-29**.

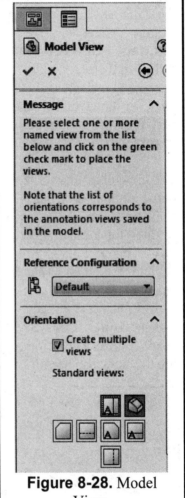

Figure 8-28. Model View.

Inserting a Section View

This next step will be covered in greater detail in Unit 9. A section view is created by doing a theoretical cut through an object and showing it in an adjacent view. The command is found under the **Insert – Drawing View – Section**. After you

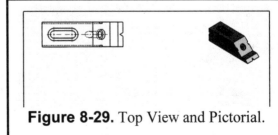

Figure 8-29. Top View and Pictorial.

have selected Section, a menu will show up where the Feature Manager Tree is on the left side of the work area. Select the Horizontal Section Icon as can be seen in **Figure 8-30**. After selecting the **Horizontal Section**, move your cursor to hover over either large arc of the slot and it will snap the cutting plane through the center of the slot. **Okay** your selection. As you drag your cursor below the Top view, a section view will follow your cursor. Place the view with adequate space somewhere to the lower left of the Title Sheet. Now **Right Click** on **Section View A-A** in the Feature Manager Tree and select **Tangent Edges**. Select **Tangent Edges Removed**. No hidden lines are to be shown in a full or half section view. If hidden lines are showing in the section view, activate the Blue dashed box around the section view and click on the Center (plane) box in the Feature Manager Tree. This will turn off the Hidden Lines. To complete the section view, **Insert - Annotations – Centerlines**.

The cross hatching of section lines is not supposed to be parallel or perpendicular to Object lines. Click on one of the section lines. This will activate a menu window that permits you to alter the settings for the hatch **(Figure 8-31)**. Uncheck the Material Hatch box. This will allow you to change the density of the hatch and the

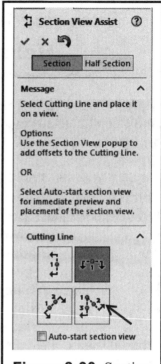

Figure 8-30. Section Cutting Line.

angle of the hatch. The section cutting line is the heaviest line used in graphics, so one additional change needs to be made. Go to **Tools – Options – Document Properties**, and under **View Labels** select **Section**. A window will appear that allows you to make changes (**See Figure 8-32**). Under **Line Style** make sure that **Phantom line** is selected and the line weight is changed to **0.0138**. When you **Okay** these changes you will notice that the cutting plane line in the top view has become a heavy black line.

Figure 8-31. Hatch Modification.

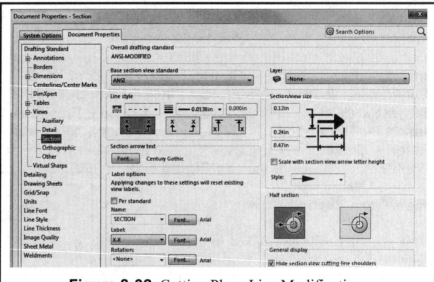

Figure 8-32. Cutting Plane Line Modification.

Inserting a Projected View

In order to assure that views are properly aligned, it is essential that they be inserted as projected views rather than placing them as regular views. You will now place a view to the right of the Section view. Go to **Insert – Drawing View – Projected View**. You will be prompted to select a drawing view from which to project. Activate the **Section View**. Then move the cursor toward the right of the section view. The right view will appear and follow the cursor. Place the view toward the right border line. Because you projected from a view that had the hidden lines removed, you will see that there are no hidden lines in the right side view. After you okay the placement go to the menu in the Feature Manager Tree and turn on the hidden lines. You should always have the hidden lines turned on except in full and half sections. As in the other views, go to **Insert – Annotations – Centerline** to complete this view.

Inserting an Auxiliary View

When you study the views placed on the titlesheet you will notice that the counterbore in the inclined surface never show as true circles. When dimensioning a drawing it is best to show this feature in a true shape view. This requires you to insert an auxiliary view. Select **Insert – Drawing View – Auxiliary**. When you do, you will get a prompt to select a Reference Edge to continue. **Pick** the **inclined edge** in the section view. This will place an arrow in the direction of sight and allow you to place the auxiliary view in direct alignment with that inclined

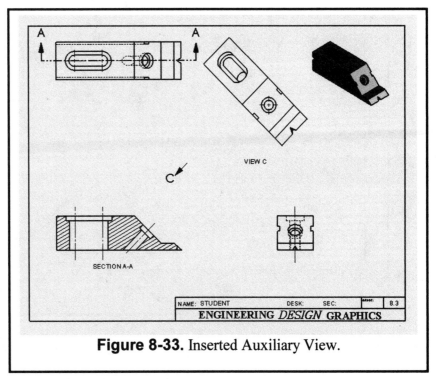

Figure 8-33. Inserted Auxiliary View.

surface. Turn on the Hidden lines. Also **Insert – Annotations – Centerlines**. This now completes the auxiliary and your drawing should be similar to **Figure 8-33**.

Inserting a Drawing Detail

Occasionally there are portions of a drawing that are too small to dimension effectively in the standard views; however, if that portion could be shown larger, it could dimension clearly. You will now **Insert – Drawing View – Detail** to show a larger view of the small V-shape of the Toe Clamp in the Top View. Once you have selected the **Detail View** command, go to the top view and draw a small circle around the V. When you move away from the V you will notice an enlarged version of that circle. Place that view at some open area of your drawing. The detail

Name and the Scale are displayed below the image. If the size is too large or too small, you can right click on the blue dashed window around the view to change the scale. Your completed drawing should look similar to **Figure 8-34**.

Now save your drawing as **TOE CLAMP PROJECTED VIEWS.slddrw** and print a copy to be turned in to your instructor.

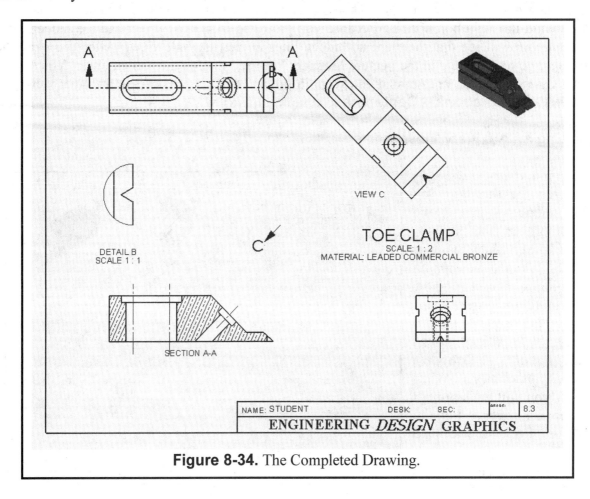

Figure 8-34. The Completed Drawing.

Exercise 8.4: Rapid Prototyping of a Solid Model Part

In this Exercise 8.3, you will build a solid model (or presumably already have one built) as assigned by your instructor. You will then save it in a file format that is the standard in the rapid prototyping industry. This file format is called stereo lithography and is abbreviated .STL. You will then send your .STL file to an available rapid prototyping machine to make a rapid physical model of your part.

SAVING THE SOLID MODEL AS A STEREO LITHOGRAPHY (.STL) FILE

You can now build your **New** solid model in SOLIDWORKS, or **Open** a model that has already been built. When you have completed the solid model, do a **Save As**, select "Save as Type" and select **.STL** Files (see **Figure 8-35**). Some example rapid physical prototypes of solid models are shown in **Figure 8-36**.

Note: If you need some ideas for solid models for this exercise, examine the accompanying Assignments 8.3.1 to 8.3.4.

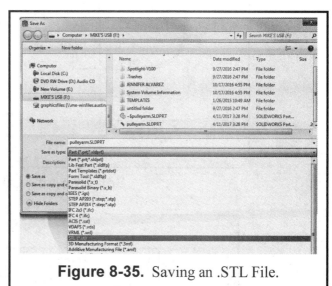

Figure 8-35. Saving an .STL File.

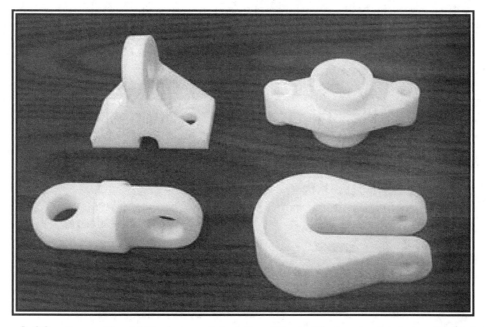

Figure 8-36. Some Rapid Prototype Models of Parts Shown in Assignments 8.3.1 to 8.3.4.

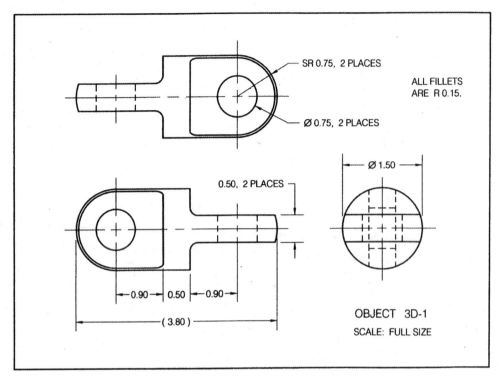

Solid Model Assignment 8.4.1

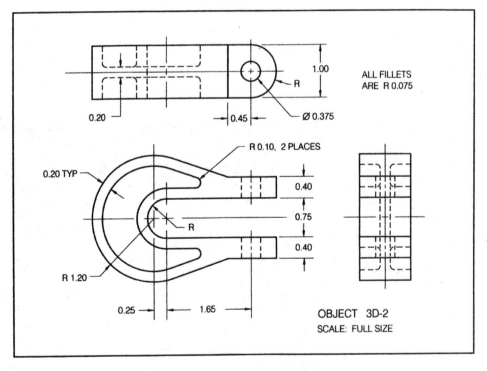

Solid Model Assignment 8.4.2

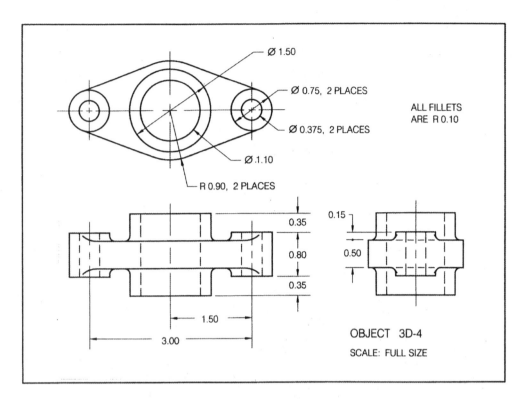

Solid Model Assignment 8.4.3

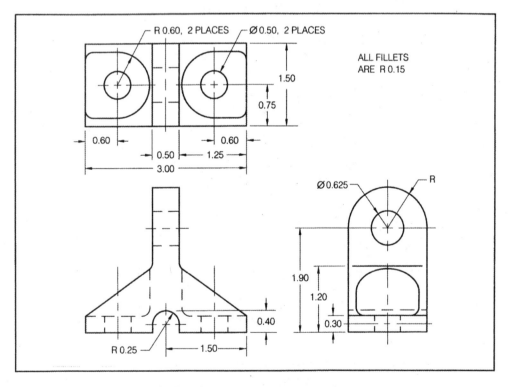

Solid Model Assignment 8.4.4

NOTES:

Computer Graphics Lab 9: Section Views in 3D and 2D

In Computer Graphics Lab 9, you will learn how to make section views from solid models. In each exercise, you will build a 3D solid model of a part. You will then use a "Display Section View" function in SOLIDWORKS that allows you to temporarily remove part of the solid model (e.g. the front half) to visualize the internal features of the 3D part. You will open a Title Block in SOLIDWORKS and drag and drop the solid model into the Title Block. SOLIDWORKS automatically will then project chosen orthographic views (front, top, and/or right sides) of the part. You can then use a function in SOLIDWORKS that allows you to create a section drawing view of the object.

VIEWING 3D SECTION VIEWS

The temporary section view of a 3D solid model can be displayed by accessing the **View** pull down menu, then select **Display** and **Section View** as shown in **Figure 9-1**. When you select the "Section View" item, a menu appears on the screen as shown in **Figure 9-2**. With this menu, you can select the cutting plane (e.g. Front plane), the position of that cutting plane relative to the origin, and which side of the model to show. You can preview the 3D section view of the model as shown in **Figure 9-3**. You can then return to the **View** pull down menu, select **Display** and deselect the **Section View** command to re-display the full 3D solid model image.

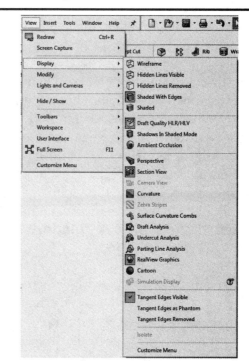

Figure 9-1. The Display Section View Command.

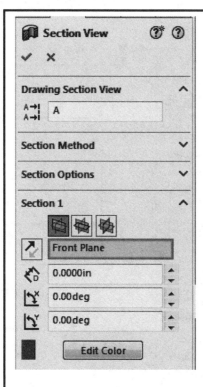

Figure 9-2. The Section View On-Screen Menu.

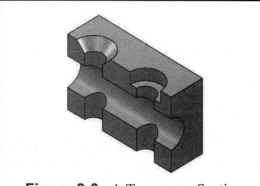

Figure 9-3. A Temporary Section View of a 3-D Solid Model.

CREATING 2D SECTION VIEWS ON A DRAWING SHEET

A 2D section view of a model can be created in a drawing file. First you must open a **TITLE BLOCK (INCHES or METRIC).drwdot** file. You start with the title block and then "drag and drop" the 3D solid model onto the drawing sheet. SOLIDWORKS is designed to automatically make a three-view orthographic layout of the model once it is dropped onto the sheet. This then becomes the starting point for making a detailed engineering drawing. From here, you can add centerlines, dimensions, notes, and title block information. One other thing you can do is to make a section view on the drawing.

You first decide which view will be the source for a section view projection. For example, the top view could be the source for projecting a full section in the front view. Select the Front view and **Delete** it. **Sketch** a **Line** that extends beyond the view by at least 0.25" to represent the cutting plane line on this source view. Then select the **Insert** pull-down menu, select **Drawing View** and then **Section**, as shown in **Figure 9-4**.

A dynamic section view now appears on the screen that you can move away from, but aligned with, the source view. For example, if the source view is the top view, then the front section view would be aligned vertically with the top view. Since SOLIDWORKS cannot anticipate which side of the source view you want to project from, you have the option to "Flip Direction" of the section view. SOLIDWORKS also adds a labeled cutting plane line (e.g. A-A) and a section view title (e.g. Section A-A).

For example, the solid model part of **Figure 9-3** was dropped into a drawing sheet. A line was drawn across the full width of the top view, and a full section view was created in the front view using the above procedure. The resultant drawing for this example is shown in **Figure 9-5**.

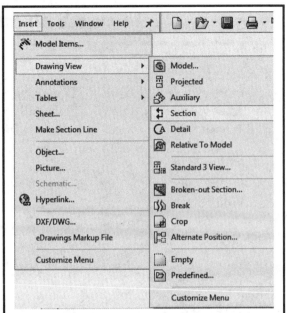

Figure 9-4. Inserting a Section View on a Drawing Sheet.

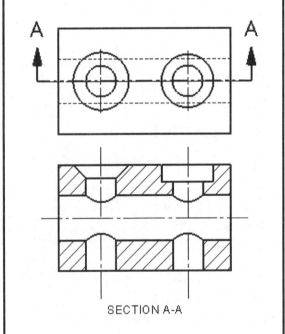

Figure 9-5. Creating a Labeled Section View on a 2-D Drawing Sheet.

Exercise 9.1: ROD BASE SECTION VIEW

In Exercise 9.1, you will build a solid model of the Rod Base using SOLIDWORKS commands that you have learned in previous labs. You will display a 3D section view of the model to visualize the internal features of the model. Next you will "drag and drop" the model onto a drawing sheet and then create a 2D section view of the part. To get started, you will first build the solid model part.

BUILDING THE ROD BASE

Open your **ANSI-INCHES.prtdot** in SOLIDWORKS. Immediately **SAVE AS – ROD BASE.sldprt**. Select the **Front** plane and start a new **Sketch**. Draw a vertical **Centerline** through the origin. Sketch the given profile in **Figure 9-6** using the **Line** tool. Apply the given dimension values using the **Dimension** tool. This will be "Sketch1" for the part.

Select the **Revolve Base** feature icon and revolve the profile into a solid using a **360** degrees angle. This forms the base part as suggested by **Figure 9-7**.

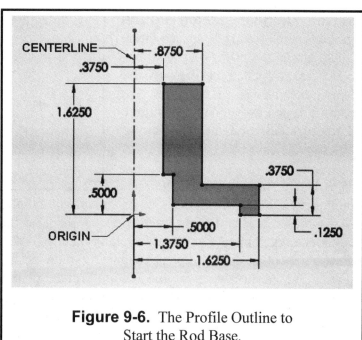

Figure 9-6. The Profile Outline to Start the Rod Base.

Now add a **Chamfer** design feature to the *top inner edge* of the through hole. Use a **Distance-Distance** parameter definition with both distance values being **0.125** inches. Next add three **Fillet** design features to the following edges.

1. *Top outer edge* of the upright cylinder.
2. *Intersection edge* between the upright cylinder and the base plate.
3. *Top outer edge* of the base plate.

Use a common fillet radius of **0.0625** inches. When you are finished with the fillets, your part should look like **Figure 9-7**.

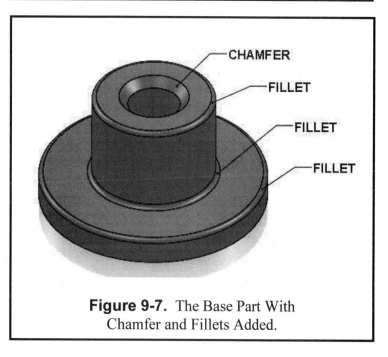

Figure 9-7. The Base Part With Chamfer and Fillets Added.

Now click on the top surface of the base plate of the Rod Base (it should turn *blue*). Also choose a **Top** view orientation and select the **Sketch** icon. Sketch a **Circle** on this base plate surface and align it with the origin. **Dimension** its diameter to be **0.2500** inches and its distance from the origin to be **1.1250** inches. Now use the **Circular Sketch Pattern** tool to create **4** equally spaced circles (total angle = 360) along this base plate, as suggested in **Figure 9-8**. Now execute an **Extruded Cut** command and use an end condition of **Through All** to create four through holes on the base plate.

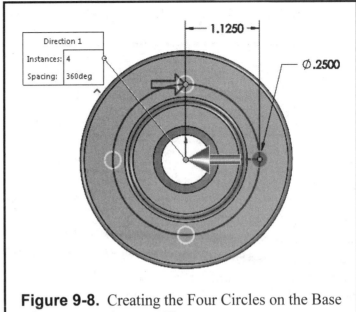

Figure 9-8. Creating the Four Circles on the Base Plate.

The final feature to add is the small pinhole that goes through one side of the upright cylinder. Select the **Right** plane in the Feature Manager and a **Right** view orientation. *Note:* The right plane is in the middle of the hollow cylinder, so you are not actually sketching on any surface of the part. Now in the **Sketch** mode, draw a **Circle** on this right plane, aligned with the origin. **Dimension** its diameter to be **0.1875** inches and its distance from the bottom surface of the base to be **1.250** inches, as shown in **Figure 9-9**. Now **Extruded Cut** the hole through the right side of the cylinder wall using an **Up to Surface** end condition in "Direction 1," where the "up to" surface selected is the outer surface of the upright cylinder. *Note:* You could also just use a **Through-All** end condition for the hole. **Right Click** on **Material (not specified)** in the Feature Manager Tree. Expand **Copper Alloys** and select **Manganese Bronze**. Click on **Apply** and **Close**. Either way, your model of the Rod Base is complete and it should look like **Figure 9-10** in an **Isometric** view. At this point you can **Save** your part in your designated folder and name it **ROD BASE**.

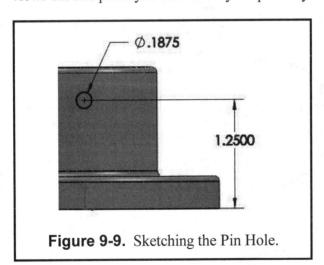

Figure 9-9. Sketching the Pin Hole.

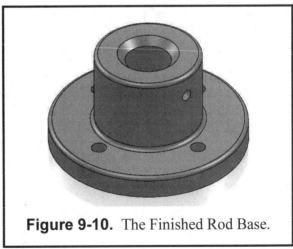

Figure 9-10. The Finished Rod Base.

MAKING A 3-D SECTION VIEW OF THE ROD BASE

You will now make a 3D section view of the Rod Base model. First select the cutting plane by clicking the **Front** plane in the Feature Manager. Next, pull down **View**, select **Display**, and then select **Section View**. The "Section View" on-screen menu now appears (refer back to **Figure 9-2**). Leave the "Section Position" at **0.0000** inches (middle of model); click the box with the arrows in it. This box is located to the left of the Front Plane box in the feature manager tree. Notice the preview arrow pointing in the back direction on the model, which is the correct direction for the section view. So then click **OK** to close the menu. You should now have a 3D full section view of the Rod Base showing a cut-away view of the part, as shown in **Figure 9-11**.

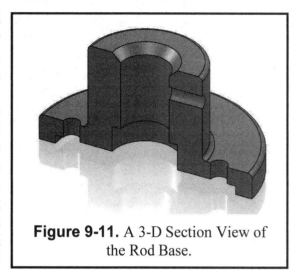

Figure 9-11. A 3-D Section View of the Rod Base.

Now return to the **View** pull down menu, re-select **Display**, and then re-select **Section View** again to turn **off** the 3D section view, or select the Section View icon. _Do not_ **Close** your solid model file yet because you will need it for the next phase of this exercise; however, **SAVE** your solid as **ROD BASE.sldprt**.

INSERTING THE ROD BASE ON A TITLE BLOCK

At this point you will make a layout of your model on a Title Block sheet. Open the file **TITLEBLOCK-INCHES.drwdot** from your folder and immediately **SAVE AS – ROD BASE.slddrw**.

Before you start the projection, there are some general settings that you need to make. Pull down **Tools** and select **Options**. Click the **Drawings, Display Style** tab. Check the dot (•) on the "Hidden lines visible" option and check the dot (•) on the tangent edges "Removed" option as shown in **Figure 9-12**. Next click the **Drawings, Area Hatch** tab. Select the **ANSI31** (Iron Brick Stone) "Pattern" option. Set the

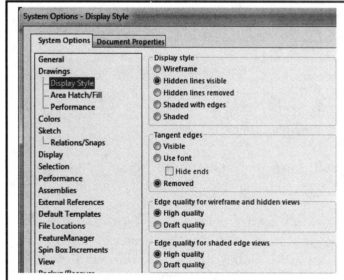

Figure 9-12. Setting Some Drawing Display Options.

"Scale" to **1.000** and the Angle" to **0** degrees, as shown in **Figure 9-13**. Also go to **Document Properties**, expand **View Label,** and select **Section**. Under **Line Style** select the **Phantom** line and change the thickness to **0.0138in**.

Now pull down **Window** and select **Tile Vertically**. You now can see the two files: the Rod Base part file and the Title Block. Activate the Title Block and go to the **Insert** pull down menu. **Select Drawing View – Model**. The name "**ROD BASE**" will appear in the **Open Documents** box. Click on the **White Circle with the Blue arrow** in it. In the orientation Box **Click** the box next to "**Create Multiple Views**." The menu, where the Feature Manager Tree normally is, will prompt you to identify the views you wish to project. Select the **Top view** and the **Isometric View**. They will display using the third angle projection you just set, as displayed in **Figure 9-14**. You may have to right click on Sheet 1 in the Feature Manager and select

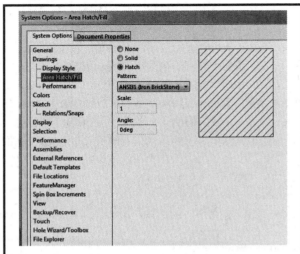

Figure 9-13. Setting Some Hatch Pattern Options.

Edit Sheet for the views to show on the Title Block. Select the **Blue Dashed box** around the **Isometric View**. Make sure that the **scale** is set to **1:2** and turn on the **shaded with edges mode**. **Right Click** in the active window, **Select Tangent Edge** and **Select Tangent Edges Visible**. Move the view into the upper right-hand corner of the title block. Now activate the blue dashed box around the top view to set its scale to **1:1** and make sure the Hidden lines are turned on. Then move the image to the upper left portion of the Title Block. **Select** the four Center Marks of the small holes and delete them. Next draw a **Circle** centered at the center of the object and through the center of one of the small holes. Turn this circle into a centerline circle by clicking the **For Construction** box. Okay your selection.

MAKING A 2D SECTION VIEW OF THE ROD BASE

The Top view is the source view to insert a full section view. You will not need to access the Rod Base model file anymore, so *maximize* the Title Block sheet by clicking its box in the upper right corner.

Pull down the **Insert** menu, select **Drawing View**, and then pick **Section**. A menu will appear with the Horizontal Section line showing as the default (See **Figure 9-15**). Move your cursor toward the center of the Top View. The cutting Plane Line will snap to the center. The Section Menu will ask you to Select OK. When you do, drag your cursor downward to the place where you want to position your Section view. Notice the vertical

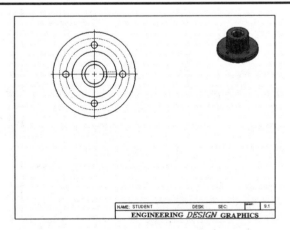

Figure 9-14. The Top View of the Rod Base.

alignment between the two views is maintained as you move the section view. Also, a cutting plane line labeled **A-A** appears where you placed the Section Line.

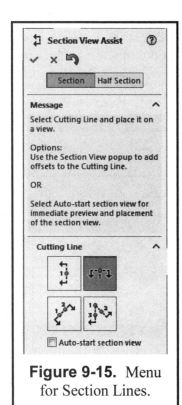

Figure 9-15. Menu
for Section Lines.

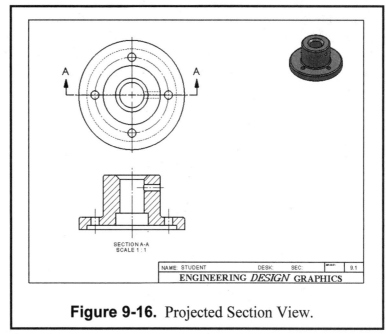

Figure 9-16. Projected Section View.

You now have a drawing with a section view that is labeled **Section A-A**, as indicated in **Figure 9-16**. If the arrows are pointing downward, check the box in the menu that will **flip direction**. If Hidden lines are showing in the Section View, they must be turned off. Finally, select **Insert – Annotations – Centerline** and check the box in front of the **Select View** to place the centerlines in the Section View.

You may want to increase the font size of the "Section A-A" label to **18 point**.

Now insert the right side view. Select **INSERT Drawing view – Projected view**. Activate the section view and you will see a frame following the cursor as you move it toward the right of the section view. When you reach the desired location, click the left mouse button to position the view. This ensures that the right side view lines up with the section view. Turn on the hidden lines in the right side view. See **Figure 9-19** for an example of the finished drawing. Use **Insert, Annotations**, and **Note** to add a **ROD BASE** title and **SCALE 1:1** to the drawing as shown in **Figure 9-18**. Use **all capitals, Arial** font for all the labels. Use **24 point** font for the "**ROD BASE**" title and **12 point** font for all the other annotations.

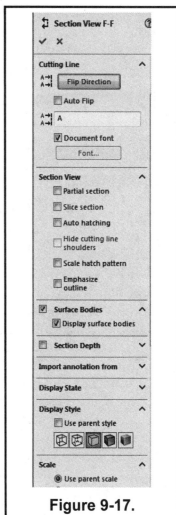

Figure 9-17.

When you are finished you should **Save** your drawing file as **ROD BASE.slddrw** in your designated folder (*note*: its extension is now **.slddrw**). A finished section view drawing, with title block and pictorial, is shown in **Figure 9-18** in a **Print Preview**. Then **Print** a hard copy of the drawing to submit to your instructor.

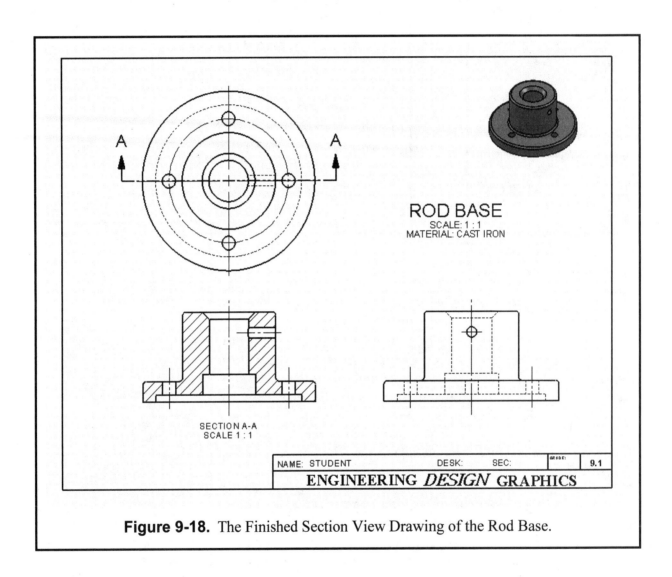

Figure 9-18. The Finished Section View Drawing of the Rod Base.

Exercise 9.2: TENSION-CABLE BRACKET SECTION VIEW

In this Exercise 9.2, you will build a solid model of the Tension Cable Bracket using SOLIDWORKS commands that you have learned in previous labs. You will display a 3-D section view of the model to visualize the internal features of the model. Next you will "drag and drop" the model onto a drawing sheet and then create a 2-D section view of the part. To get started, you will first build the solid model part.

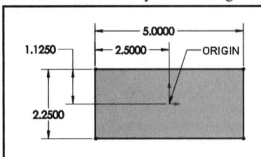

Figure 9-19. The Initial Outline to Start the Part.

BUILDING THE TENSION CABLE BRACKET

Open your **ANSI-INCHES.prtdot** in SOLIDWORKS. Immediately **SAVE AS – TENSION-CABLE BRACKET.sldprt**. Go to **Tools – Options – Document Properties – Units** and set your decimal places to **Three**. Select the **Top** plane and start a new **Sketch**. Draw a **Rectangle** centered at the origin as indicated in **Figure 9-19**. Apply the given dimension values using the **Dimension** tool. This will be "Sketch1" for the part. Select the **Extrude Base** feature icon and extrude it **0.250** inches upward from the plane. This is the base plate.

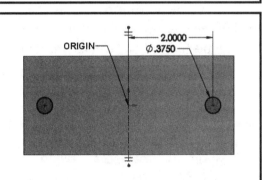

Figure 9-20. Sketching the Two Plate Holes.

Now click on the top surface of this base plate (it turns *blue*). **Sketch** the right side **Circle** on the top plate and **Dimension** it as shown in **Figure 9-20**. Add a horizontal relationship between the hole and the origin. Draw a vertical **Centerline** and **Mirror** the circle to the left side. Next, **Extruded Cut** the holes **Through All** with a **Draft** angle of **15** degrees, as suggested in **Figure 9-21**. This results in the two base plate tapered holes.

Now return to the top surface of the base plate and start another **Sketch** on it. Draw another **Circle**, centered at the *origin* and with a *diameter* of **1.750** inches. Execute an **Extruded Boss** command with the following data:

    "Direction 1" (up)   = **Blind 0.875** inches
    "Direction 2" (down)   = **Blind 0.875** inches

You now have a boss going in each direction.

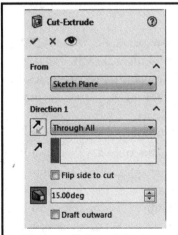

Figure 9-21. The Data for the Holes.

Now click on the top surface of the top boss (it turns *blue*) and **Sketch** a **Circle** on it. Center the circle at the *origin* and give it a *diameter* of **1.000** inch. Next **Extruded Cut** the hole **Through All** with a **Taper** angle of **7.5** degrees. This makes a big tapered hole through the part as shown as a wireframe in **Figure 9-22**.

You will now add some triangular supports to the boss feature. These will be offset from the center, so you first need to add a new sketching plane. Click on the **Front** plane in the Feature Manager. Pull down **Insert**, select **Reference Geometry**, and then select **Plane**. Make the new **Plane1** a distance **0.750** inches from and in front of the Front plane, and then click the (√) button. The new Plane1 is now added to the Feature Manager. Click on this **Plane1** and make a **Sketch**. Sketch three **Lines** that form a triangle and **Dimension** them as shown in **Figure 9-23**. Draw a vertical **Centerline** through the origin, select **Mirror**, then select the three lines. This gives you a mirrored image of the triangle.

Now execute an **Extruded-Boss** for this sketch, using a **Blind** distance of **0.250** inches towards the direction of the origin (into the object). You now have two triangular supports on the front side of the object. While holding down the CTRL button select the **Front** plane and the last extrusion in the Feature Manager and select the **Mirror Feature** icon, then click the (√) button. You now have a part as shown in **Figure 9-24**.

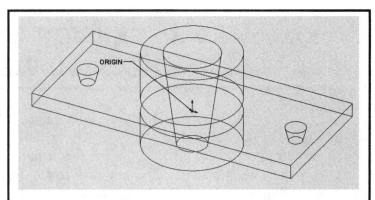

Figure 9-22. Wireframe Model of Cable Bracket at This Stage in the Building Process.

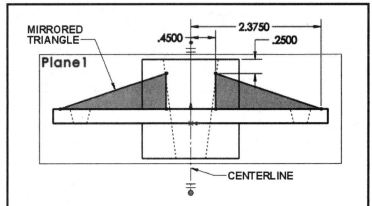

Figure 9-23. Sketching and Mirroring a Triangular Support on Plane1.

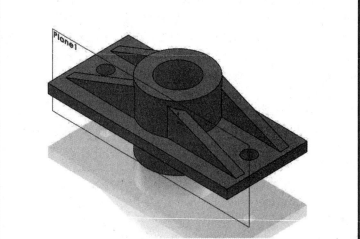

Figure 9-24. The Four Triangular Supports Added.

You now can complete the Tension Cable Bracket by adding some design features. First add a **Chamfer** feature to the inner hole edge on top of the upright boss. Define the chamfer with a **Distance-Distance** parameter that have both distances equal to **0.125** inches. Next add **Fillet** features to a number of edges. Make them all to have a uniform *radius* of **0.0825** inches. Pick the following edges to fillet with this command:

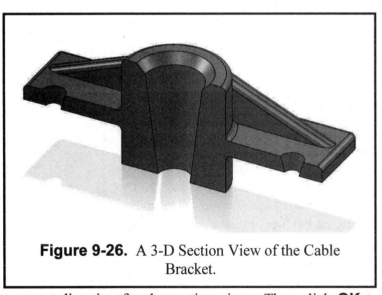

Figure 9-25. The Finished Tension Cable Bracket after Adding the Chamfer and Fillets.

- Outer top edge of upright boss.
- All intersections of the upright boss with the base plate, both above and below the base plate.
- The top outer edges of the base plate.
- The top edges of the four triangular supports.
- The vertical edges of the rectangular base.

When you are finished, you will have a finished part as shown in **Figure 9-25** in a **Trimetric** view. At this point you may wish to **Save** your part in your designated folder and name it **TENSION CABLE BRACKET.sldprt**.

MAKING A 3D SECTION VIEW OF THE TENSION CABLE BRACKET

You will now make a 3D section view of the Cable Bracket model. First select the cutting plane by clicking the **Front** plane in the Feature Manager. Next, pull down **View**, select **Display**, and then select **Section View**. The "Section View" on-screen menu now appears (refer back to **Figure 9-2**). Leave the "Section Position" at **0.000** inches (middle of model), check (√) **on** the "Flip the Side to View," and then press the **Display** button on the menu. Notice the preview arrow pointing in the back

Figure 9-26. A 3-D Section View of the Cable Bracket.

direction on the model, which is the correct direction for the section view. Then click **OK** to close the menu. You should now have a 3-D full section view of the Cable Bracket showing a cut-away view of the part, as shown in **Figure 9-26**.

Now return to the **View** pull down menu, re-select **Display**, and then re-select **Section View** again to turn **off** the 3D section view. *Do not* **Close** your solid model file yet because you will need it for the next phase of this exercise.

INSERTING THE CABLE TENSION BRACKET ON A TITLE SHEET

At this point you will make a three-view layout of your model on a Title Block sheet. Open the file **TITLEBLOCK-INCHES.drwdot** from your folder and immediately **SAVE AS – CABLE TENSION BRACKET.slddrw**.

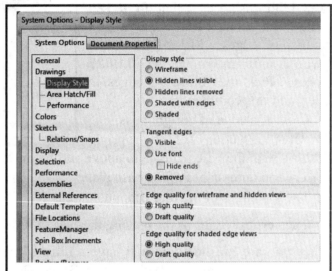

Figure 9-27. Setting Some Drawing Display Options.

Before you start the projection, there are some settings that you need to make. Pull down **Tools** and select **Options**. Click the **Drawings**, **Display Style** tab. Check the dot (•) on the "Hidden lines visible" option and check the dot (•) on the tangent edges "Removed" option as shown in **Figure 9-27**. Next, click the **Drawings**, **Area Hatch/Fill** tab. Select the **ANSI31** (Iron Brick Stone) "Pattern" option. Set the "Scale" to **1.000** and the Angle" to **0** degrees. Also go to **Document Properties**, expand **View Label,** and select **Section.** Under **Line Style** change the thickness to **0.0138in.**

Now pull down **Window** and select **Tile Vertically**. You now can see the two files: the Cable Bracket part file and the Title Block. Activate the Title Block and go to the **Insert** pull down menu. **Select Drawing View – Model**. The name **"CABLE BRACKET"** will appear in the **Open Documents** box. Click on the **Blue Circle with the White arrow** in it. In the orientation Box **Click** the box next to **"Create Multiple Views."** The menu, where the Feature Manager Tree normally is, will prompt you to identify the views you wish to project. Select the **Top view** and the **Isometric View**. Select the **Blue Dashed box** around the **Isometric View**. Make sure that the **Scale** is set to **1:2** and turn on the **Shaded Mode**. **Right Click** in the active window, and move the view into the upper right-hand corner of the title block. Now activate the **Blue dashed**

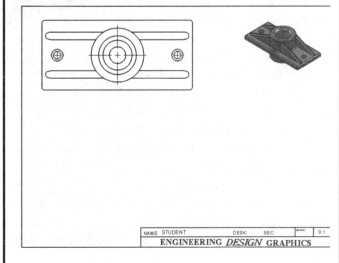

Figure 9-28. The Top View of the Cable Bracket.

box around the **Top view** to set its **Scale to 1:1** and make sure the Hidden lines are turned on. Then move the image to the upper left portion of the Title Block.

MAKING A 2D SECTION VIEW OF THE CABLE BRACKET

You will only need the Title Block Sheet file now, so *maximize* it by clicking its box in the upper right corner. Pull down the **Insert** menu, select **Drawing View**, and then pick **Section**. A menu will appear with the Horizontal Section line showing as the default **(See Figure 9-29)**. Move your cursor to hover over the circles in the **Top View**. Then move toward the center that highlights. The Cutting Plane Line will snap to the center. The Section Menu will ask you to **Select OK**. When you do, drag your cursor downward to the place where you want to position your Section view. Notice the vertical alignment between the two views is maintained as you move the section view. Also a cutting plane line labeled **A-A** appears where you placed the Section Line. The cutting plane arrows should point in the correct direction as you drag the section view into position. If the arrows are pointed downward, select the flip direction box. Find a good position and click the **LMB** to fix the section view in that position. You now have a drawing with a section view that is labeled **Section A-A**, as indicated in **Figure 9-30**.

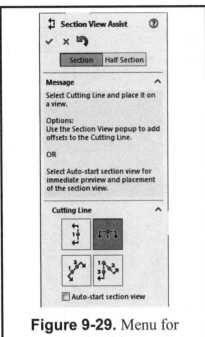

Figure 9-29. Menu for Section Lines.

The section view drawing may not be perfect. For example, the centerlines in the section view may be missing. You can **Select – Insert, Annotations, Centerlines** to insert them. You may want to increase the font size of the "Section A-A" label to **18 point**.

Next, activate the section view and select **Insert – Drawing View – Projected view**. Move the cursor to the right of the section view and when it is in the desired position, click the **RMB**. Make sure that you turn on the hidden lines in the right side view.

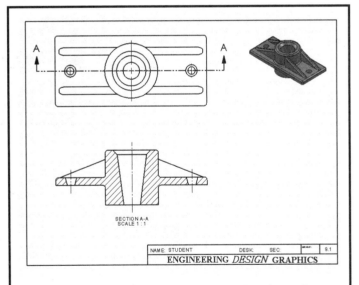

Figure 9-30. Projecting a Front Section View.

Use **Insert, Annotations,** and **Note** to add a **TENSION CABLE BRACKET** title and **SCALE 1:1** to the drawing as shown in **Figure 9-31**. Use *caps*, **Arial** font for all the labels. Use **24-point** font for the "Tension Cable Bracket" title and **12-point** font for all the other annotations.

When you are finished you should **Save** your drawing file as **TENSION CABLE BRACKET.slddrw** in your designated folder. A finished section view drawing, with title block and pictorial, is shown in **Figure 9-31**. **Print** a hard copy of the drawing to submit to your instructor.

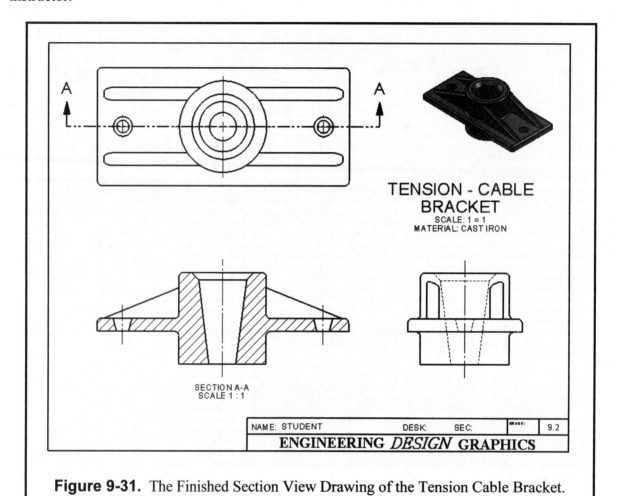

Figure 9-31. The Finished Section View Drawing of the Tension Cable Bracket.

Exercise 9.3: MILLING END ADAPTER
Section Views

In Exercise 9.3, you will build a solid model of the Milling End Adapter using SOLIDWORKS commands that you have learned in previous labs. You will display a 3-D section view of the model to visualize the internal features of the model. Next you will "insert" the model onto a drawing sheet and then create a 2-D section view of the part. To get started, you will first build the solid model part.

BUILDING THE MILLING END ADAPTER

Open your **ANSI-INCHES.prtdot** in SOLIDWORKS. Immediately **SAVE AS – MILLING END ADAPTER.sldprt**. Also set the "Major Grid Spacing" to **1** inch and "Minor Lines per Major" to **16**. Make sure the "Display Grid" is also set to **on**. The grid will not show until you activate the sketch command. Select the **Front** plane and start a new **Sketch**. If you do not get 16 grids per inch, **zoom in until they appear**. Draw a horizontal **Centerline** through the origin. Sketch the given profile in **Figure 9-32** using the **Line** tool. Follow the grid pattern in the figure in accordance with the previous settings to make your sketch profile.

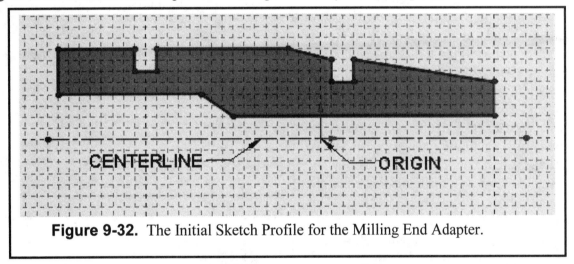

Figure 9-32. The Initial Sketch Profile for the Milling End Adapter.

Select the **Revolve Base** feature icon and revolve the profile into a solid using a **360** degrees angle. This forms the base part. Now add two **Chamfer** design features to the end edges of the part. Use a **Distance-Distance** parameter definition, with both distance values being **0.0625** inches. At this stage, you should have a revolved solid part with the ends chamfered, as shown in **Figure 9-33** in a **Trimetric** view.

Figure 9-33. The Revolved Part with Chamfers.

You now need to create two screw holes on the round outer surfaces of the Milling End Adapter. SOLIDWORKS needs a flat plane to sketch on, so first select the **Top** plane in the Feature Manager. Next pull down **Insert**, select **Reference Geometry**, and then **Plane**. The "Plane" menu appears. Key-in the plane distance of **0.5000** inches *above* the Top plane as indicated in **Figure 9-34**. Then click the green (√) mark to close the menu.

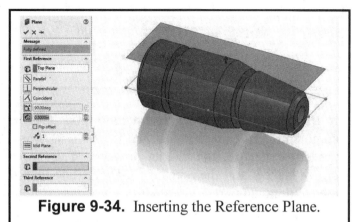

Figure 9-34. Inserting the Reference Plane.

You now have a "Plane1" in the Feature Manager Tree structure. Click on that **Plane1** and enter the **Sketch** mode. Also select a **Top** view orientation to better see the Plane1 sketch plane. Sketch two **Circles** on the plane and **Dimension** them as shown in **Figure 9-35**. The diameters are both **0.1250** inches. The first circle center is **0.8125** inches from the **Origin** of the Milling End Adapter, and the second circle center is **0.3750** inches to the left of the first circle center. To totally fix the geometry, add a horizontal relation to the Origin.

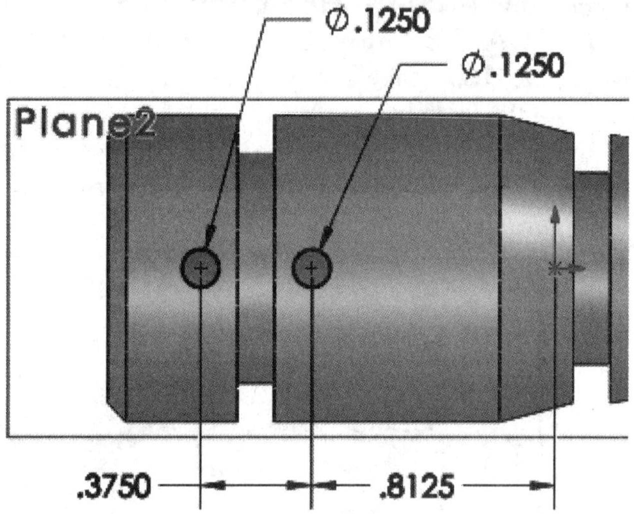

Now **Extruded Cut** the two holes down through the cylinder wall using an **Up to Surface** end condition in "Direction 1," where the "up to" surface selected is the *inner* hollow surface of the shaft. This results in the holes only going through the top part. *Note*: You could also just use a **Blind** end condition for the holes with the downward distance being around **0.500** inches. Either way, your model of the Milling End Adapter is complete and it should look like **Figure 9-36** in a **Trimetric** view. **Right Click** on **Material (not specified)** in the Feature Manager Tree. Expand **Copper Alloys** and select **Manganese Bronze**. Click on **Apply** and **Close**. At this point **Save** your part in your designated folder and name it **MILLING END ADAPTER.sldprt**.

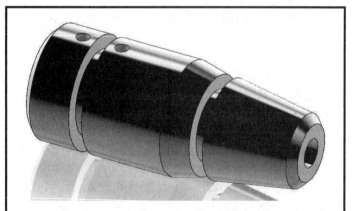

Figure 9-35. Adding Two Circles for the Screw Holes.

Figure 9-36. Finished Milling End Adapter Part.

MAKING A 3D SECTION VIEW OF THE MILLING END ADAPTER

You will now make a 3D section view of the Milling End Adapter model. First select the cutting plane by clicking the **Front** plane in the Feature Manager. Next, pull down **View**, select **Display**, and then select **Section View**. The "Section View" on-screen menu now appears (refer back to **Figure 9-2**). Leave the "Section Position" at **0.0000** inches (middle of model), check (√) **on** the "Flip the Side to View," and then press the **Display** button on the menu. Notice the preview arrow pointing in the back direction on the model, which is the correct direction for

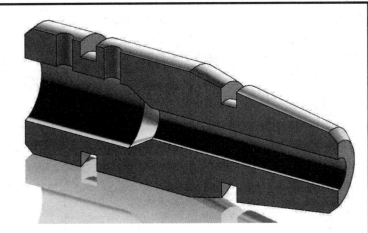

Figure 9-37. A 3-D Section View of the Milling End Adapter.

the section view. So then click **OK** to close the menu. You should now have a 3D full section view of the Milling End Adapter showing a cut-away view of the part, as shown in **Figure 9-37**.

Now return to the **View** pull down menu, re-select **Display**, and then re-select **Section View** again to turn **off** the 3D section view, or select the Section icon 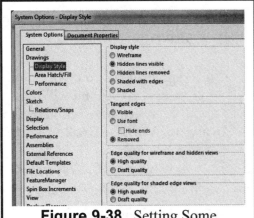. _Do not_ **Close** your solid model file yet because you will need it for the next phase of this exercise.

INSERTING THE MILLING END ADAPTER ONTO A TITLE BLOCK

At this point you will make a three-view layout of your model on a Title Block sheet. Open the file **TITLEBLOCK-INCHES.drwdot** from your folder and immediately **SAVE AS – MILLING END ADAPTER.slddrw**.

Before you start the projection, there are some general settings that you need to make. Pull down **Tools** and select **Options**. Click the **Drawings, Display Style** tab. Check the dot (•) on the "Hidden lines visible" option and check the dot (•) on the tangent edges "Removed" option as shown in **Figure 9-38**. In the Feature Manager Tree, **Right Click** on "**Material <not specified>**. Select **Edit Material** and expand **Copper Alloys**. Select **Manganese Bronze** for the material to be assigned to the Milling End Adapter.

Figure 9-38. Setting Some Drawing Display Options.

Now pull down **Window** and select **Tile Vertically**. You now can see the two files: the Milling End Adapter part file and the Title Block. Activate the Title Block and go to the **Insert** pull down menu. **Select Drawing View – Model**. The name **"Milling End Bracket"** will appear in the **Open Documents** box. Click on the **Blue Circle with the White arrow** in it. In the orientation Box **Click** the box next to **"Create Multiple Views."** The menu, where the Feature Manager Tree normally is, will prompt you to identify the views you wish to project.

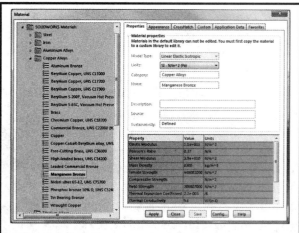

Figure 9-39. Applying Material to a Part.

Select the **Top view Front View, Right Side** and **Pictorial View**. When you **Okay** your choices the views will appear on the Title Sheet. Now *maximize* your Drawing Sheet file to full screen to see it better. They will display using the third angle projection you just set, as displayed in **Figure 9-40**. However, you may discover that the drawing scale is too small for this application. So return to **Sheet1, Properties** menu and set the "Scale" to **2:1** to get a bigger layout like the one shown **Figure 9-40**. You may also activate the views and change the custom scale of the view in the Feature Manager column. **Insert – Annotations – Centerlines** in the Three views.

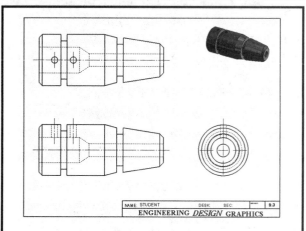

Figure 9-40. The Three Views of the End Adapter, Using a 2:1 Scale.

MAKING A 2D SECTION VIEW OF THE END ADAPTER

For this application, you will use all three views to make a layout and will show the front view with a broken-out section. Select **Insert – Drawing View - Broken-Out Section** icon.

You will be prompted to "draw a closed spline to continue section creation." To define the broken-out section area, **Sketch** a somewhat arbitrarily enclosed **Spline** on the front view of the end adapter, as suggested in **Figure 9-41**. You can *grab* the nodes around the

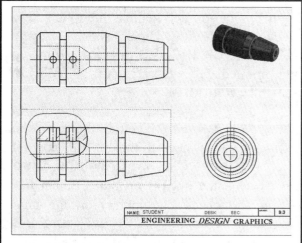

Figure 9-41. Sketching an Enclosed Spline on the Front View for the Broken-Out Section.

completed spline and drag them in or out of the loop to set your area until it is acceptable.

In the "Broken-out Section" menu in the Feature Manager area, as shown in **Figure 9-42**, click on (√) in the **Preview** box. You now get a preview of the cutting plane in both the top and right side views, including view direction arrows. On the "Broken-out Section" menu, key in the "Distance" for the cutting plane to be **0.5000** inches. This places the cut right through the middle of the end adapter. Then click the green (√) button to close it.

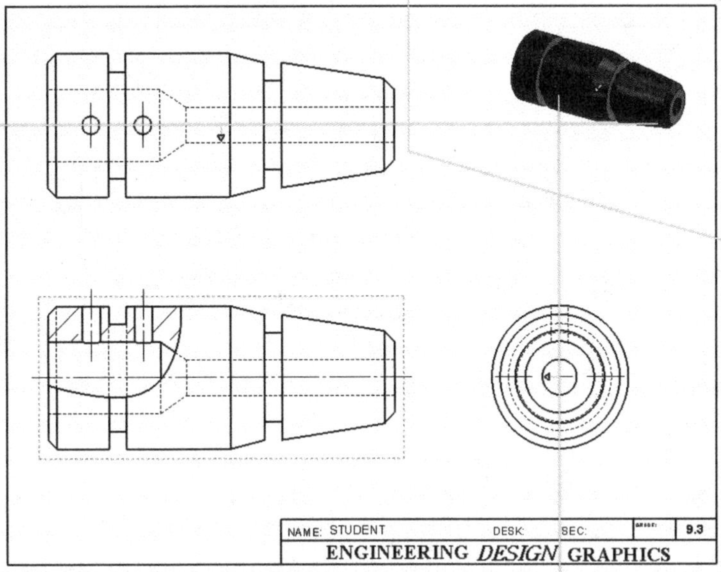

Note: You can use a **Rectangle** instead of a Spline in this **Broken-out Section** command to make a *Half*

Figure 9-42. Inserting the Broken-out Section View and Previewing It.

Section View of a part using this same approach.

You now have a Broken-Out Section view of the Milling End Adapter in the front view, and also top and right side views, as suggested in the final **Figure 9-43**. However, it may not be a perfect drawing. **Insert – Annotations - Centerlines** in the front and top views if they are missing. You can **Insert, Annotations**, and select **Center Mark**. Then click on the perimeter of the circular features to add center marks (i.e., centerlines) to circular views. You may want to add a label below the front view that says **BROKEN-OUT SECTION** using **14**-point font size.

Use **Insert, Annotations**, and **Note** to add the **MILLING END ADAPTER** title and **SCALE 2:1** and **MATERIAL: BRASS** to the drawing as shown in **Figure 9-44**. Use *caps*, **Arial** font for all the labels, **20-point** font for the "Milling End Adapter" title, and **12-point** font for the other annotations.

When you are finished you should **Save** your drawing file as **MILLING END ADAPTER.slddrw** in your designated folder (note: its extension is now **.slddrw**). A finished broken-out section view drawing, with title block and border, is shown in **Figure 9-43** in a **Print Preview**. Then **Print** a hard copy of the drawing to submit to your instructor.

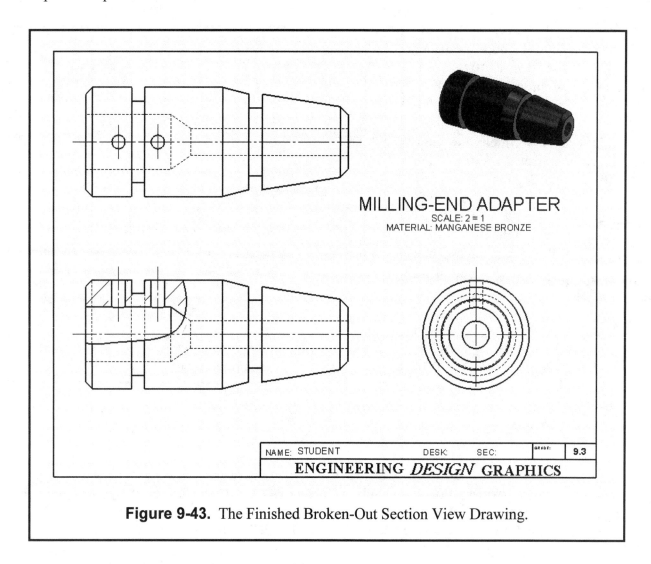

MILLING-END ADAPTER
SCALE: 2 = 1
MATERIAL: MANGANESE BRONZE

| NAME: STUDENT | DESK: | SEC: | GRADE: | 9.3 |

ENGINEERING *DESIGN* GRAPHICS

Figure 9-43. The Finished Broken-Out Section View Drawing.

Exercise 9.4: PLASTIC REVOLVING BALL ASSEMBLY Section View

In Exercise 9.4, you will build the three parts of the Plastic Revolving Ball assembly using SOLIDWORKS commands that you have learned in previous labs. You will display a 3D section view of the model to visualize the internal features of the model. Next you will "drag and drop" the assembly model onto a title block and then create a 2D section view of the assembly. To get started, build the solid model of each part and then create an assembly of the three parts.

Building the Ball

Open **ANSI-INCHES.prtdot** from your folder. Immediately **SAVE AS – PLASTIC BALL.sldprt**. Select the **Material Icon** (**Figure 9-44**) in the Feature Manager. Expand the **Plastics** option and select **PE High Density** and under the **Appearance Tab** select the **"Use Material Color."** Select the **Front** plane and start a new sketch. Draw a horizontal **Centerline** through the **Origin**. **Sketch** the given profile in **Figure 9-44** and apply the given dimensions. Add a relation between the center of the small arc and the Origin to be Vertical.

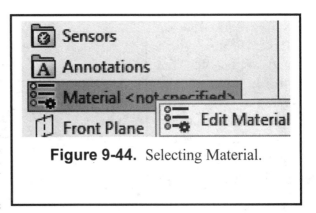

Figure 9-44. Selecting Material.

Select the **Revolve Base** feature icon and revolve the profile into a solid using a **360**-degree angle. This creates the Plastic Ball. Right click on **Material (not specified)** under **Plastics** and select **PF**. Select **APPLY** and then **CLOSE**. Your object should look like **Figure 9-46**.

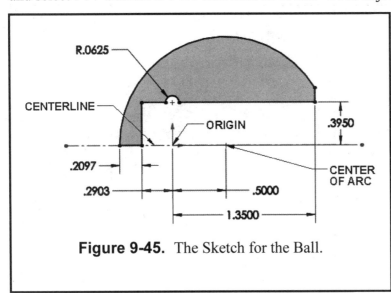

Figure 9-45. The Sketch for the Ball.

Figure 9-46. The Finished Plastic Ball.

Building the Steel Shaft

Open **ANSI-INCHES.prtdot** from your folder. Immediately **SAVE AS – STEEL SHAFT.sldprt**. Select the **Material Icon** (**Figure 9-48**) in the Feature Manager. Expand the **Steel** option and select **AISI 1020** and under the **Appearance Tab** de-select the **"Use Material Color."** Select the **Front** plane and start a new sketch. Draw a horizontal **Centerline** through the **Origin**. **Sketch** the given profile in **Figure 9-48** and apply the given dimensions.

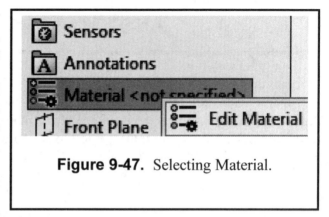

Figure 9-47. Selecting Material.

Select the Revolve Base Feature icon and **Revolve** the profile into a solid using a **360**-degree angle. This creates the Steel Shaft. Select the small end of the shaft and start a new sketch. **Sketch** the

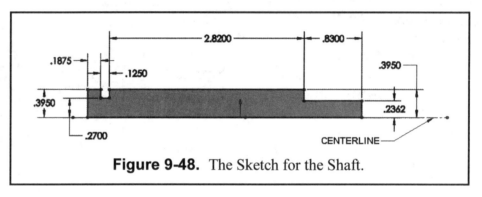

Figure 9-48. The Sketch for the Shaft.

Hexagon as shown in **Figure 9-49** and do a **Cut Extrude - Blind** of **0.375** inches. Add a **0.0375**-inch **Chamfer** on both ends of the shaft. Right click on **Material (not specified)** and select **AISI Annealed Stainless Steel (SS)**. Select **APPLY** and then **CLOSE**. Your object should look like **Figure 9-50**. **Save** your second part as **STEEL SHAFT.sldprt**.

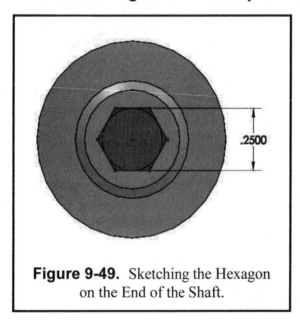

Figure 9-49. Sketching the Hexagon on the End of the Shaft.

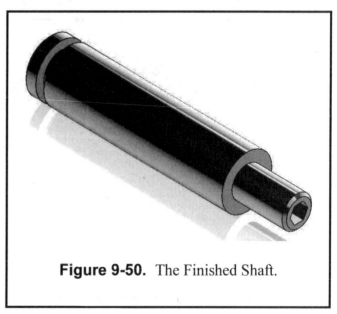

Figure 9-50. The Finished Shaft.

Building the Snap Ring

Open **ANSI-INCHES.prtdot** from your folder. Immediately **SAVE AS – SNAP RING.sldprt**. Select the **Material Icon** (**Figure 9-52**) in the Feature Manager. Expand the **Steel** option and select **Alloy Steel** and under the **Appearance Tab** de-select the **"Use Material Color."** Select the **Front** plane and start a new sketch. Draw a horizontal **Centerline** through the **Origin**. **Sketch** the given profile in **Figure 9-52** and apply the given dimensions. **Constrain** the sketch so it is **centered above the origin**.

Select the Revolve Base Feature icon and **Revolve** the profile into a solid using a **360**-degree angle. This creates the Snap Ring.

Select the right plane and start a new sketch. **Sketch** the 2-D profile as shown in **Figure 9-53** and do a **Cut Extrude – Through all in both directions**.

Your object should look like **Figure 9-54**. Change the shading **Color** to **Golden Yellow**, and then **Save** your third part as **SNAP RING.sldprt**.

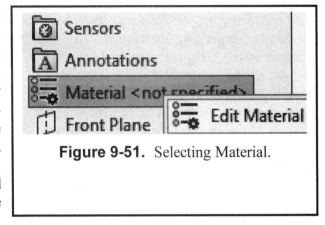

Figure 9-51. Selecting Material.

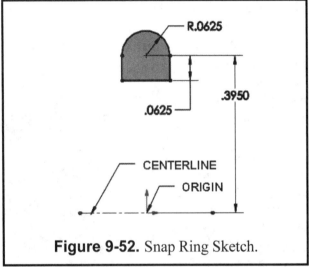

Figure 9-52. Snap Ring Sketch.

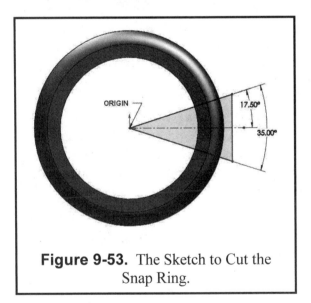

Figure 9-53. The Sketch to Cut the Snap Ring.

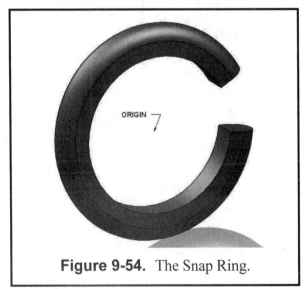

Figure 9-54. The Snap Ring.

Building the Assembly

With the three part files open, go to **File** and select **New, Assembly**. In the pull-down menu "Windows" select **Tile Vertically**. Activate the Assembly window. Pull down the **View** menu and turn on the **Origins**. First drag the Snap-Ring to the **Origin** (when you see a double set of arrows) of the Assembly as the base part. Next select the Steel-Shaft and drag it into the assembly **BUT NOT TO THE ORIGIN.** Add the following **Mates** to the parts. The steel shaft should be **Concentric** with the snap-ring for the first mate and then make one of the vertical surfaces of the snap ring and the corresponding vertical surface of the ring groove of the shaft to be **Coincident**. This will position the snap ring in the groove of the steel shaft. When this is completed, drag the Plastic-Revolving Ball to the **Origin** (Take the cursor arrow to the origin. When you see a double set of arrows, release the mouse button) of the Assembly. **Save** the assembly to your folder as **PLASTIC REVOLVING BALL.sldasm**.

MAKING A 3-D SECTION VIEW OF THE PLASTIC REVOLVING BALL ASSEMBLY

You will now make a 3-D section view of the Plastic Revolving Ball Assembly. First select the cutting plane by clicking the **Front** plane in the Feature Manager. Next pull down **View**, select **Display**, and then select **Section View**. The "Section View" on-screen menu now appears (refer back to **Figure 9-2**). Leave the "Section Position" at **0.0000** inches (middle of model),

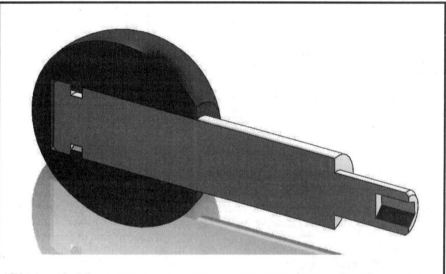

Figure 9-55. A 3D Section View of the Revolving Ball Assembly.

check (√) **on** the "Flip the Side to View," and then press the **Display** button on the menu. Notice the preview arrow pointing in the back direction on the model, which is the correct direction for the section view. Then click **OK** to close the menu. You should now have a 3D full section view of the Plastic Revolving Ball Assembly, showing a cut-away view of the part, as shown in **Figure 9-55**.

Now select the Section icon ![icon] to return to the original view. _Do not_ **Close** your assembly model file yet because you will need it for the next phase of this exercise. You may want to close the individual parts at this time.

INSERTING THE REVOLVING BALL ASSEMBLY ONTO A TITLE BLOCK

Before you start the projection, there are some settings that you need to make. Pull down **Tools** and select **Options**. Click the **Drawings, Display Type** tab. Check the dot (•) on the "Hidden lines visible" option and check the dot (•) on the tangent edges "Removed" option as shown in **Figure 9-56**.

Go to **Tools Options** and select **Document Properties,** expand **View Label,** and select **Section**. Under **Line Style** change the thickness to **0.0138in**.

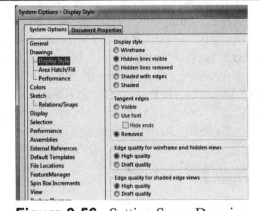

Figure 9-56. Setting Some Drawing Display Options.

Now pull down **Window** and select **Tile Vertically**. You now can see the two files: the Cable Bracket part file and the Title Block. Activate the Title Block and go to the **Insert** pull down menu. **Select Drawing View – Model**. The name "**PLASTIC REVOLVING BALL**" will appear in the **Open Documents** box. Click on the **White Circle with the Blue arrow** in it. In the orientation box, **Click** the box next to "**Create Multiple Views**." The menu, where the Feature Manager Tree normally is, will prompt you to identify the views you wish to project. Select the **Top view** and the **Isometric View**. Select the **Blue Dashed box** around the **Isometric View**.

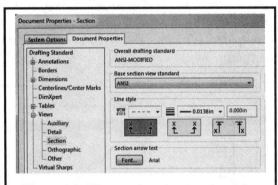

Figure 9-57. Setting the Section Line Width.

Make sure that the **Scale** is set to **1:2** and turn on the **Shaded Mode**. **Right Click** in the active window; move the view into the upper right-hand corner of the title block. Now activate the **Blue dashed box** around the **Top view** to set its **Scale to 1:1** and make sure the Hidden lines are turned on. Then move the image to the upper left portion of the Title Block.

Pull down the **Insert** menu, select **Drawing View**, and then pick **Section**. A menu will appear with the Horizontal Section line showing as the default (See **Figure 9-58)**. Move your cursor toward the center of the Top View. The Cutting Plane Line will snap to the center. The Section Menu will ask you to **Select OK**. When you do, drag your cursor downward to the place where you want to position your Section view. Notice the vertical alignment between the two views is maintained as you move the section view. Also, a cutting plane line labeled **A-A** appears where you placed the section view. The cutting plane arrows should point in the correct direction as you drag the section view into position. If the arrows

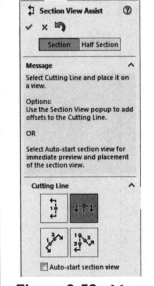

Figure 9-58. Menu for Section Lines.

are pointed downward, select the flip direction box. Find a good position and click the **LMB** to fix the section view in that position. You now have a drawing with a section view that is labeled **Section A-A**, as indicated in **Figure 9-59**. Once the section view is placed, you will notice that the section lines may not be at different angles and the scale of the section lines is not appropriate. If you **Right Click** on the section lines, a window appears that lets you edit the **Crosshatch Properties**. Make changes in the Hatch Pattern Scale and the Hatch Pattern Angle to change the density of the Hatch and their directions. Next activate the section view and go to **Insert - Drawing View – Projected View** to place the right side view. You will notice that centerlines and hidden lines are missing in the right side view. Turn on the hidden lines, then select **Insert, Annotations, Center Mark**. SOLIDWORKS will prompt you to "select a circular edge for center mark insertion." Select the largest circle in the right side view and the center mark will be added.

Use **24 point** font for the "**PLASTIC REVOLVING BALL**" title and **12 point** font for all the other annotations.

When finished you should **Save** your drawing file as **PLASTIC REVOLVING BALL.slddrw** in your designated folder. A finished section view drawing, with title block and border, is shown in **Figure 9-59** in a **Print Preview**. Then **Print** a hard copy of the drawing to submit to your instructor.

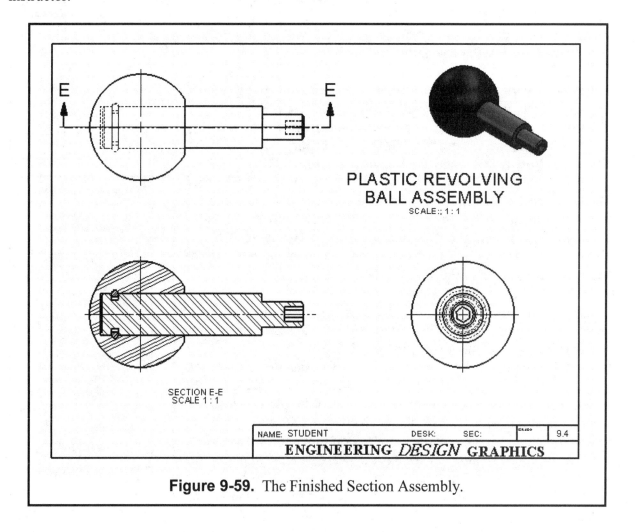

Figure 9-59. The Finished Section Assembly.

SUPPLEMENTARY EXERCISE 9-5: CLAMPING BLOCK

Build a full size solid model of the figure below. Insert it on a Title Block and provide a **Full Section** in the place of the front view. Insert a small pictorial of the object in the upper right hand corner of the sheet. Provide the proper Titles, Scales and other pertinent notes. Also go to **Document Properties,** expand **View Label,** and select **Section.** Under **Line Style** change the thickness to **0.0138in.**

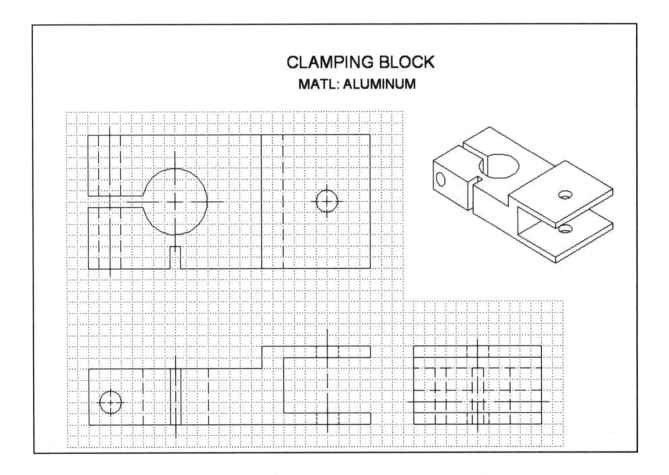

CLAMPING BLOCK
MATL: ALUMINUM

ASSUME THE GRID DIVISIONS TO BE 3 mm.

SUPPLEMENTARY EXERCISE 9-6: TWO WAY BENCH BLOCK

Build a full size solid model of the figure below. Insert it on a Title Block. Go to **Document Properties**, expand **View Label**, and select **Section.** Under **Line Style** change the thickness to **0.0138in**. After deleting the front and the right side view, draw an offset line through the two counterbored holes and the center hole. Next go to **Insert – Drawing View – Section View** and place it where the front view used to be. Continue by Inserting – Drawing View – Projected view to replace the right side view. Finally insert a small Isometric View in the upper right hand corner and supply the Title and Scale as in previous exercises.

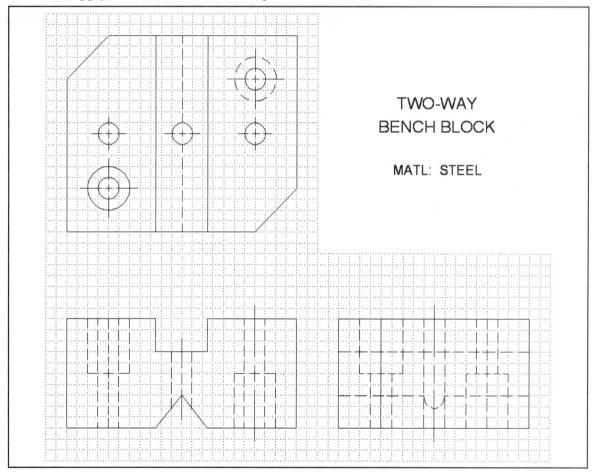

TWO-WAY
BENCH BLOCK

MATL: STEEL

ASSUME THE GRID DIVISIONS TO BE 0.125 INCHES.

SUPPLEMENTARY EXERCISE 9-7: ELECTRICAL CONTACT PLATE

Build a full size solid model of the figure below. Insert it on a Title Block. Go to **Document Properties,** expand **View Label,** and select **Section.** Under **Line Style** change the thickness to **0.0138in.** After deleting the front and the right side view, draw an offset line through one of the small holes on the left end, then through the slot with the two holes in the center and through one of the slots on the right hand end. Next go to **Insert – Drawing View – Section View** and place it where the front view used to be. Continue by **Inserting – Drawing View – Projected view** to replace the right side view. Finally insert a small Isometric View in the upper right hand corner and supply the Title and Scale as in previous exercises.

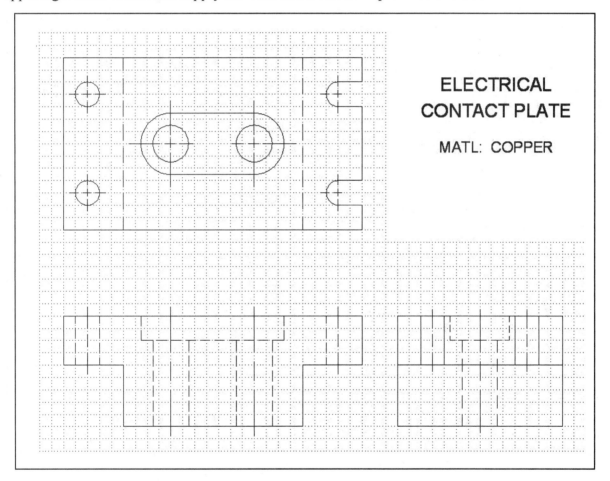

ELECTRICAL
CONTACT PLATE

MATL: COPPER

ASSUME THE GRID DIVISIONS TO BE 0.125 INCHES.

SUPPLEMENTARY EXERCISE 9-8: DISC ASSEMBLY

Build full sized solid models of the **Disk Assembly Pieces** below. Build a fully constrained Assembly of the parts and insert it on a Title Block. Go to **Document Properties,** expand **View Label,** and select **Section.** Under **Line Style** change the thickness to **0.0138in**. Provide a **Full Section** in the place of the front view. Finally, insert a small Isometric View in the upper right hand corner of the unexploded assembly in the upper right hand corner of your Title Block. Provide the proper Titles, Scales and other pertinent notes.

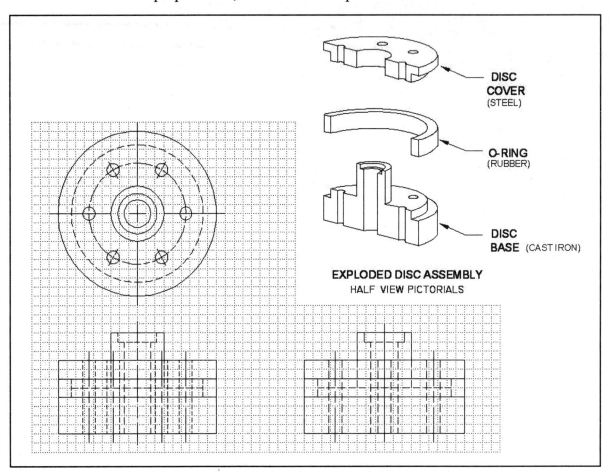

ASSUME THE GRID DIVISIONS TO BE 0.25 INCHES.

Computer Graphics Intro to Lab 10: Setting Up Templates for Three View Drawings

Up to this point you have used the Title Block as a drawing sheet on which you place your solid models; however, the Title Sheet can store many of the settings that are used multiple times. For instance, in the last unit you were required to change the thickness of the cutting plane line for every exercise. Had you saved the setting for the line thickness on your template, that line thickness would have been created automatically.

FORMATTING THE TITLEBLOCK-INCHES.drwdot

Open the **TITLEBLOCK-INCHES.drwdot** drawing sheet. In the **Tools** pull down menu, you can select the **Options** menu to get all the tab settings for your file. Click on the **Drawings, Display Style** tab, set the projection line options, such as dot (•) **on** the "Hidden lines visible" and dot (•) **on** the "Tangent Edges Removed" as shown in **Figure 10-1**.

SETTING THE DIMENSIONING VARIABLES

The next step is to set some of the dimensioning parameters using the pull-down **Tools, Options**.

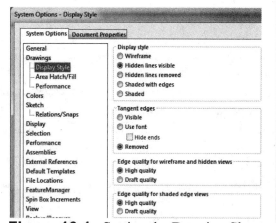

Figure 10-1. Setting the Drawing Sheet Default Display Types Projections.

Select the **Document Properties** tab and select the **Dimension** menu. This opens the menu, as shown in **Figure 10-2**. While in this menu make the following changes.
Bent Leaders – change to **.125** inches.
Arrows – change values to **.042, .125, and .25** respectively. At the same time, select **Font** and set the dimension **Font, Font Style,** and **Height** as follows:

Font = **Arial**
Font Style = **Regular**
Points = **12**

Select **OK**.

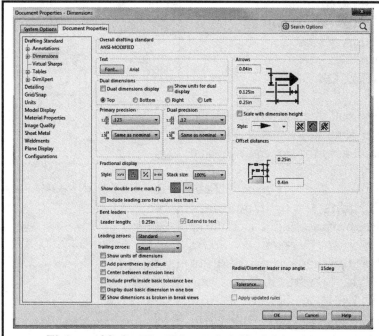

Figure 10-2. The Dimension Variables Menu.

The last step for formatting the template is to change the line thickness for the cutting plane line for section views. Pull down the **Tools Tab, Options, Document Properties** and select the **+** next to **Views**. Select **Section** and change the **Line Style** to **Phantom** and the thickness to **0.0138in**. Now save this template with all of these settings.

FORMATTING THE TITLEBLOCK-METRIC.drwdot

Open the **TITLEBLOCK-METRIC.drwdot** drawing sheet. In the **Tools** pull down menu, you can select the **Options** menu to get all the tab settings for your file. Click on the **Drawings, Display Style** tab, set the projection line options, such as dot (•) **on** the "Hidden lines visible" and dot (•) **on** the "Tangent Edges Removed" as shown in **Figure 10 -3**.

SETTING THE DIMENSIONING VARIABLES

The next step is to set some of the dimensioning parameters using the pull-down **Tools, Options**. Select the **Document Properties** tab and select the **Dimension** menu. This opens the menu, as shown in **Figure 10-4**. While in this menu make the following changes.
Bent Leaders – change to **3mm**.
Arrows – change values to **1mm, 3mm, and 6mm** respectively. At the same time, select **Font** and set the dimension **Font, Font Style**, and **Height** as follows:

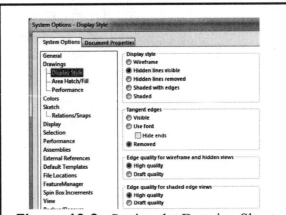

Figure 10-3. Setting the Drawing Sheet Default Display Types Projections.

Font = **Arial**
Font Style = **Regular**
Points = **12**

Select **OK**.

The last step for formatting the template is to change the line thickness for the cutting plane line for section views. Pull down the **Tools Tab, Options, Document Properties** and select the **+** next to **Views**. Select **Section** and change the **Line Style** to **Phantom** and the thickness to **.35mm**. Now save this template with all of these settings.

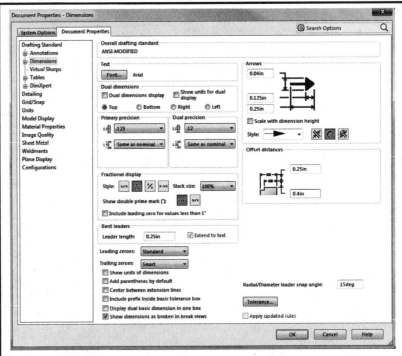

Figure 10-4. The Dimension Variables Menu.

Computer Graphics Lab 10: Generating and Dimensioning Three-View Drawings

In Computer Graphics Lab 10, you will learn how to project three-view orthographic drawings from solid models. In each exercise, you will build a 3D solid model of a part. You will then use a projection function in SOLIDWORKS to make a three-view (front, top, and right side) drawing of the part. You will completely dimension the drawing using SOLIDWORKS commands. You will finish the drawing by adding centerlines, annotations, on a Title Block. The following introduction will get you oriented to making a drawing using SOLIDWORKS.

INSERTING A DRAWING SHEET FOR THE PROJECTION

You start the three-view layout of your model on a Title Block drawing sheet. You pull down **File**, select **Open**, and select the **TITLEBLOCK-INCHES.drwdot** drawing sheet that was made earlier. The Drawing Sheet is now on the screen as shown in **Figure 10-5**.

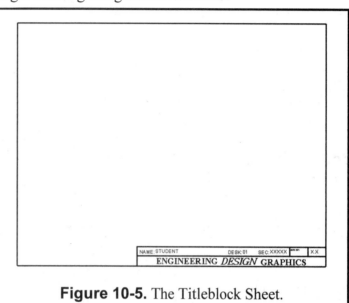

Figure 10-5. The Titleblock Sheet.

SETTING THE DRAWING SHEET DISPLAY TYPE

Before you start the projection, there are some general drawing sheet settings that you need to set. In the **Tools** pull down menu, you can select the **Options** menu to get all the tab settings for your file. Several of them are pertinent to drawing projections. If you click on the **Drawings**, **Display Style** tab, you can set the projection line options, such as dot (•) **on** the "Hidden lines visible" option and dot (•) **on** the "Tangent Edges Removed" option as shown in **Figure 10-6**.

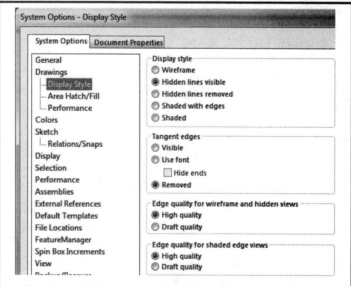

Figure 10-6. Setting the Drawing Sheet Default Display Types for the Projection.

PROJECTING THE THREE VIEWS

When you have both your model and drawing sheet active, you can pull down **Window** and select **Tile Vertically**. Here you can see the two files: the part file and the newly created blank Drawing Sheet file. Go to the **Insert** pull down menu. **Select Drawing View – Model**. The name of the model will appear in the **Open Documents** box. Click on the **White Circle with the Blue arrow** in it. In the Orientation Box **Click** the box next to "**Create Multiple Views.**" The menu, where the Feature Manager Tree normally is, will prompt you to identify the views you wish to project. Select the

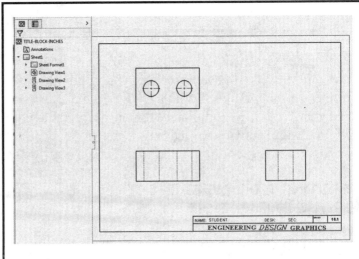

Figure 10-7. A Three-View Projection of a Part.

Top, Front, and Right Side views. Figure 10-7 is an example of such a three-view projection. Notice that the three views are arranged properly and are aligned both vertically and horizontally. You can move them around on the sheet and this alignment will remain intact.

SETTING THE DIMENSIONING VARIABLES

You may first want to set some of the dimensioning parameters using the pull-down **Tools, Options** selection. There you can select the **Document Properties** tab and select the **Dimension** menu. This opens the menu, as shown in **Figure 10-8**. Here you can set the dimension parameters such as having parentheses around a dimension and whether to have the dimension arrows inside or outside the extension lines. You will also have control of the arrowhead proportions. These settings will depend on the application and will be set later in the specific exercises for this Computer Graphics Lab 10.

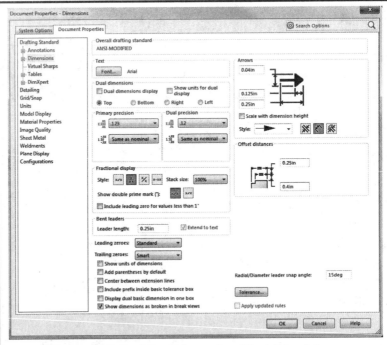

Figure 10-8. The Dimension Variables Menu.

While at the **Dimensions** tab select **Font** on the Document Properties menu (see **Figure 10-8**). Click on Font to get the menu shown in **Figure 10-9**. Here you can set the dimension **Font**, **Font Style**, and **Height** (in either units or points).

Font = **Arial**
Font Style = **Regular**
Points = **12**

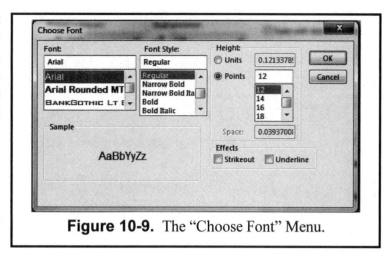

Figure 10-9. The "Choose Font" Menu.

DIMENSIONING THE DRAWING

You can now use the **Dimension** tool to add the dimensions to your three views. The new software has added an icon when you select the second dimension point. The symbol is ⬤. Whichever portion of the circular icon you choose is where the dimension will be placed automatically. By choosing this icon the spacing from the drawing view is automatic as well as the spacing between consecutive dimensions. You can also use the **Centerline** sketch tool to add any missing centerlines. An example three-view drawing, with dimensions and centerlines added, is shown in **Figure 10-10**.

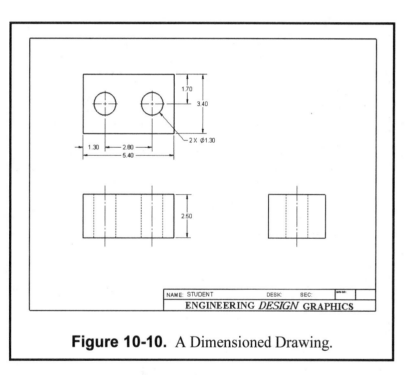

Figure 10-10. A Dimensioned Drawing.

INSERT A SHADED PICTORIAL

Go to the **Insert** Tab and Select **Drawing View**. A pop-up menu appears as shown in **Figure 10-11**. One of the selections allows you to insert a **Drawing View** of various types, such as **Model**. If you select this option, you can insert a pictorial view of the part. To facilitate this, a "Model View" menu appears in the Feature Manager Tree area.

The "Model View" menu is shown in **Figure 10-12**. Here you can set the display orientation (e.g. Isometric), the display style (e.g. Shaded), and the scale of the pictorial. Then move the cursor over to the upper right corner and insert the pictorial. The drawing can now be finished by **Insert, Annotations, Note**, as shown in **Figure 10-13**.

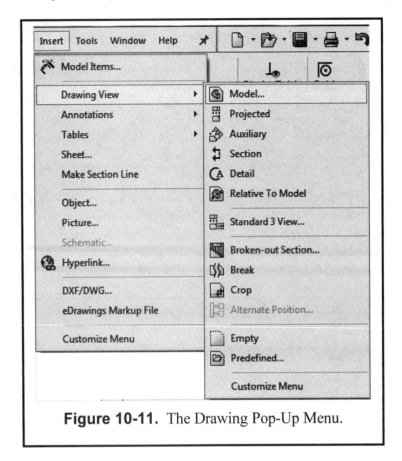

Figure 10-11. The Drawing Pop-Up Menu.

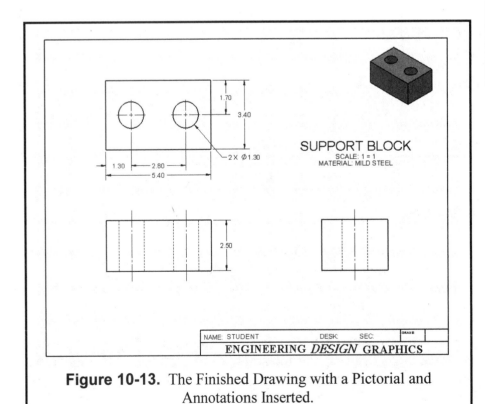

Figure 10-12.
Model View Menu.

Figure 10-13. The Finished Drawing with a Pictorial and
Annotations Inserted.

Exercise 10.1: GUIDE BLOCK Drawing

In Exercise 10.1, you will build a solid model of the Guide Block using SOLIDWORKS commands that you have learned in previous labs. Next, you will insert the model onto a drawing sheet and then create a three-view orthographic drawing of the part. You will add dimensions and complete the drawing with annotations and a title.

BUILDING THE GUIDE BLOCK SOLID MODEL

Open **ANSI-INCHES.prtdot** and immediately **SAVE AS – GUIDE BLOCK.sldprt**. Select the **TOP PLANE** and use a **Top** view orientation. Use **Inch** "Units" to **3** decimal places. In the **Sketch** mode, use the **Line** tool to draw the outline as shown in **Figure 10-14**. Use the **Dimension** tool to apply the given dimensions. Then **Extrude** the profile *upward* from the plane using a **Blind** end condition of **.75** inches.

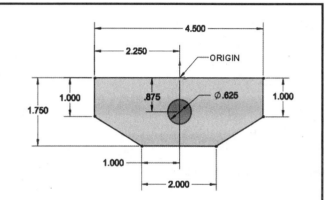

Figure 10-14. The Beginning Sketch for the Guide Block.

Change to a **Back** view orientation and click on the back surface of the model (it turns *blue*). **Sketch** a **Rectangle** and **Dimension** it as shown in **Figure 10-15**. Add a collinear relation between the edge of the base and the rectangle. Draw a vertical **Centerline** through the origin and **Mirror** the rectangle to the other side. Then in an **Isometric** view, **Extrude** this Sketch2 *forward* using a **Blind** end condition of **1.00** inches.

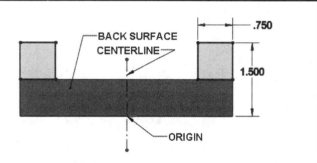

Figure 10-15. Sketching on the Back Surface.

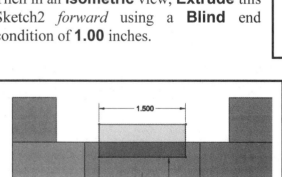

Figure 10-16. Sketching the Slot on the Front Surface.

The final feature for the Guide Block is a guide slot that goes all the way through from front to back. In a **Front** view orientation, click on the *middle* front upright surface (it turns blue). **Sketch** a **Rectangle** and **Dimension** it as shown in **Figure 10-16**. Then use **Extruded Cut** to cut the sketch using a **Through All** condition in the *backward* direction. **Right Click** on **Material (not specified)** in the Feature Manager Tree. Expand **Steel** and select **AISI 1020**. Click on **Apply** and **Close**.

The Guide Block solid model is now complete, as shown in **Figure 10-17** in a **Trimetric** view. Now **Save** your part as **GUIDE BLOCK.sldprt** in your designated folder, but *do not close* your file. You will now insert a three-view projection onto a Title Block.

PROJECTING THREE VIEWS

Pull down **File** and select the **TITLEBLOCK-INCHES.drwdot** Drawing Sheet. Immediately **Save as GUIDE BLOCK.slddrw**.

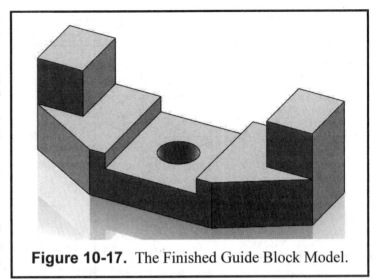

Figure 10-17. The Finished Guide Block Model.

Before you start the projection, there are some other general drawing sheet settings that you need to make. Pull down the **Tools** menu and select **Options**. Click the **System Options – Drawings - Display Style** tab. Check the dot (•) on the "Hidden lines visible" option and check the dot (•) on the tangent edges "Removed" option.

Now pull down **Window** and select **Tile Vertically**. You now can see the two files: the Guide Block part file and the newly created blank Drawing Sheet file. Go to the **Insert** pull down menu. **Select Drawing View – Model**. The name of the model will appear in the **Open Documents** box. Click on the **White Circle with the Blue arrow** in it. In the Orientation Box **Click** the box next to **Create Multiple Views**. The menu, where the Feature Manager Tree normally is, will prompt you to identify the views you wish to project. Select the **Top, Front, and Right Side views**. Notice that the views are arranged properly and are aligned vertically and horizontally, **(Figure 10-18)**.

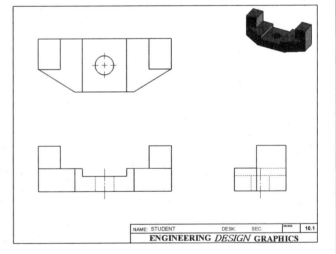

Figure 10-18. Projecting Three Views of the Guide Block onto a Drawing Sheet.

SETTING THE DIMENSION VARIABLES

You are now ready to start dimensioning the Guide Block. But first you should set some general dimensioning variables. Pull-down **Tools**, and select **Options**. Then pick the **Document Properties** tab and select the **Dimensions** menu. This opens the menu, as shown earlier in **Figure 10-9, page 10-5**. Click **off** the "Add parentheses by default" variable so that the dimensions do not have parentheses around them. Click on the **Font** tab on the **Dimensions** menu (see **Figure 10-5, page 10-5**). Here you can set the following values:

Font = **Arial**
Font Style = **Regular**
Height = **12 Points**

While in this menu make the following changes.
Bent Leaders – change to **.125** inches.
Arrows – change values to **.042, .125, and .25** respectively. Then click **OK** to close the "Dimension" menu. See **Figure 10-19**.

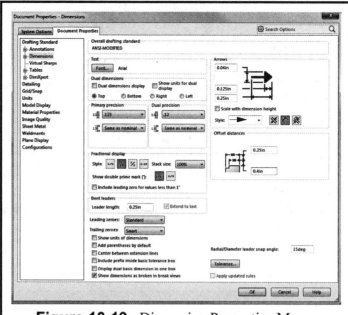

Figure 10-19. Dimension Properties Menu.

DIMENSIONING THE DRAWING

You will start dimensioning the front orthographic view. Click on the **Front** view and a cyan border appears around it. You can **Move** the border over to the right a little to make room for the dimensions if needed. Now click the **Dimension** tool icon and start picking the edges to dimension. Start with the left upright width. Select the **yellow half circle** and the **.75** value will be placed as shown in **Figure 10-20**.

Continue dimensioning some more widths (**.750**, **1.50**, and **.750**) as shown in **Figure 10-15**. Next, dimension the **.250** height of the slot. You can move this 0.25 height value outside the dimension line, but the arrows are still crowded inside the

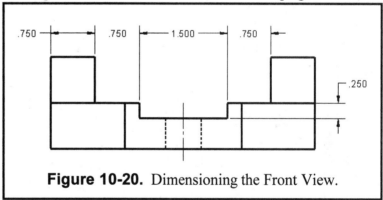

Figure 10-20. Dimensioning the Front View.

extension lines. Activate the dimension (it will turn cyan). You will notice circular dots on the arrowheads. If you click on these dots you will notice that the arrowheads will move either to the outside or to the inside as shown in **Figure 10-20**.

Now click on the **Top** orthographic view border, and **Move** it up and away from the front view, if needed, to make more room for the dimensions. Now **Dimension** the given values as indicated in **Figure 10-21**. In this case, you can pick the end *points* or *corners* for the length of the dimension (instead of picking an actual line length). Use the **RMB Properties** approach to change the default dimension settings, if needed. Next, **Dimension** the **hole and its location** as well as the **Right Side** orthographic view on your own (*hint*: there are two missing height dimensions on the right side view; see **Figure 10-22**).

Now right click the mouse (**RMB**) on the drawing sheet and the pop-up menu (**Figure 10-11**) appears. Select **Drawing View**, then **Model**. On the "Model View" menu (**Figure 10-12**) select **Trimetric** orientation, **Shaded** style, and **Scale** of **1:2**. Move the cursor to the upper right corner and insert the pictorial. Finish the drawing with **Insert, Annotations, Note** for adding a **GUIDE BLOCK** label at **24 points** and a **SCALE 1:1** and **Material Note** both at **12 points**. Use the appropriate font style and text height for the labels. Your drawing should now be complete, as suggested in **Figure 10-22**, except for the location of the hole with its diameter and the two height dimensions in the right side view. Now **Save** your drawing and name it **GUIDE BLOCK.slddrw**.

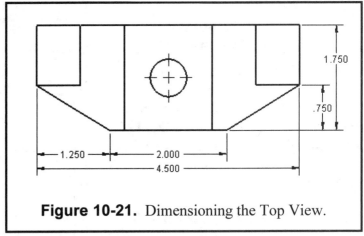

Figure 10-21. Dimensioning the Top View.

Then **Print** a hard copy to submit to your instructor.

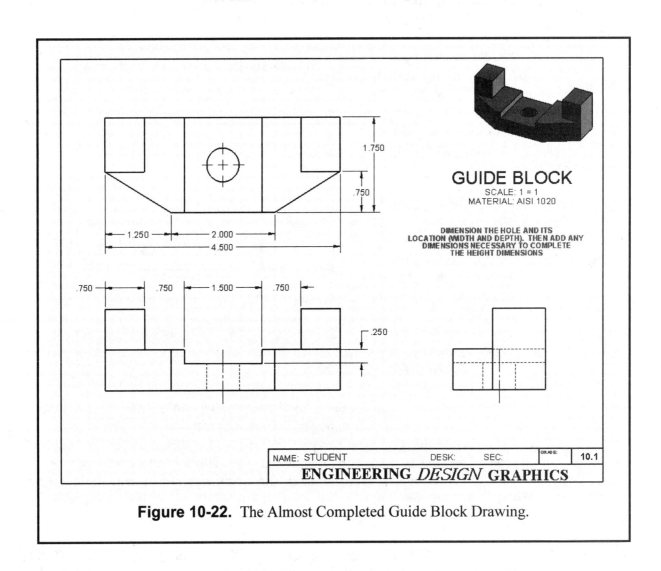

Figure 10-22. The Almost Completed Guide Block Drawing.

Exercise 10.2: PIPE JOINT Drawing

In this Exercise 10.2, you will build a solid model of the Pipe Joint using SOLIDWORKS commands that you have learned in previous labs. This exercise will be executed using millimeter units. Next, you will insert the model onto a drawing sheet and then create a three-view (top front and right side) orthographic drawing of the part. You will add dimensions and complete the drawing with centerlines, annotations, and a small pictorial on a title block.

BUILDING THE PIPE JOINT SOLID MODEL

Open the **ANSI-METRIC.prtdot**. Do a **Save As PIPE JOINT.sldprt**. Select the **Top** plane and use a **Top** view orientation. In the **Sketch** mode, use the **Circle** tool to draw two circles as shown in **Figure 10-23**. The first circle is centered at the origin; the other circle is to the right. **Dimension** the *diameters* as shown (ϕ**56** mm and ϕ**28** mm). **Dimension** the two circles **48** mm apart. Now **Add** a **Horizontal Relation** to the centers of the two circles so that they are aligned. Next, draw two **Lines** that are approximately tangent to both circles on top and bottom as shown in **Figure 10-19**. Then **Add** a **Tangent Relation** to the circle and line in *four places* as shown. Also add a vertical **Centerline** that goes through the origin and extends beyond the big circle's perimeter.

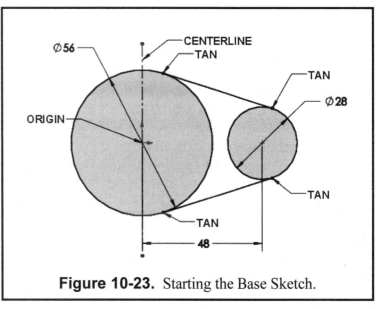

Figure 10-23. Starting the Base Sketch.

Now use the **Trim** tool to eliminate the left side of the small circle. Next **Trim** the left side of the large circle where it intersects the centerline. Then **Trim** the right side of the big circle where it is tangent to the top

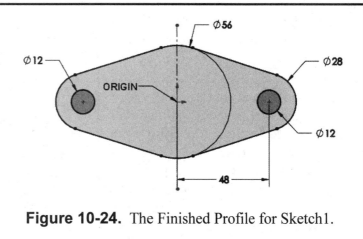

Figure 10-24. The Finished Profile for Sketch1.

and bottom lines. **Draw** a circle concentric to the small arcs with a **Diameter of 12 millimeters**. Now **Mirror** the remaining sketch (there are six pieces) about the vertical centerline to get the finished Sketch1 profile as shown in **Figure 10-24**.

Now **Extrude** this Sketch1 *upward* using a **Blind** end condition of **16** mm. This yields the base plate for the Pipe Joint, as suggested in **Figure 10-25**.

Click on the top surface of this base plate (it turns *blue*) and **Sketch** a **Circle** on the right end. This new circle should have a *diameter* of **18** mm and should have a **Concentric Relation** with the outer round edge of the base plate. Repeat this **18** mm **Circle** on the other side and also add a **Concentric Relation** with the other outer round edge. Then **Extruded Cut** this Sketch 2 *downward* **6mm** to make a counterbored hole on each end of the base plate, as shown in **Figure 10-25**.

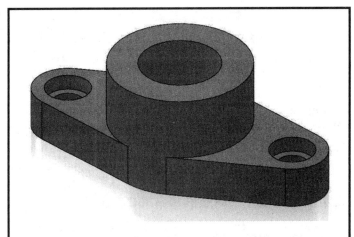

Figure 10-25. The Pipe Joint Model After:
a. The Base Plate is Extruded Upward,
b. The Side Bolt Holes are counterbored,
c. The Big Center Boss is Added, and
d. The Hole is Cut Through the Boss and Base.

Return to the top surface of this base plate (it turns *blue*) and **Sketch** a **Circle** centered above the *origin* and with a *diameter* of **56** mm (same as the previous big circle of **Figure 10-23**). Then **Extruded Boss** this new circle *upward* to a **Blind** distance of **24** mm. Finally, click on the top surface of this new Boss and **Sketch** a **Circle** centered above the *origin* and with a *diameter* of **32** mm. Then **Extruded Cut** this circle *downward* **Through All** to make a large hole that goes all the way through the boss and base plate. Your part should now look like **Figure 10.25**.

The final step to complete the Pipe Joint is to add fillets to some edges. Select the **Fillet** feature icon and then, one-by-one, click the following lines: the tangent edges on the top of the base plate (4 places), the rounded ends on the top of the base plate (2 places), the top edge of the large boss and the intersection edges of the boss with the top of the base plate (2 places). Set the fillet

radius to **3** mm and then click the check mark (√) to close the "Fillet" menu. Add a 3mm chamfer to the large hole on the top of boss. Now right click on **Material (not specified)** and change the material to **Wrought Stainless Steel**. Click on **Apply** and **Close**. The Pipe Joint model is now finished, as shown in **Figure 10-26** in a **Trimetric** view. You may now wish to **Save** your part as **PIPE JOINT.sldprt** in your designated folder, but *do not Close* your file. You can now make a three-view projection.

Figure 10-26. The Finished Pipe Joint Model with Fillets.

PROJECTING THREE VIEWS

Now **Open** your **TITLEBLOCK-METRIC.drwdot**. Go to **Save AS** and save this sheet as **PIPE JOINT.slddrw**.

Before you start the projection, there are some other general drawing sheet settings that you need to make (refer back to **Figure 10-3**). Pull down the **Tools** menu and select **Options**. Click the **System Options – Drawings - Display Style** tab. Check the dot (•) on the "Hidden lines visible" option and check the dot (•) on the tangent edges "Removed" option. Also, on the **Document Properties**, **Detailing** tab, set "Dimensioning Standard" to **ANSI**. Click **OK** to close the menu.

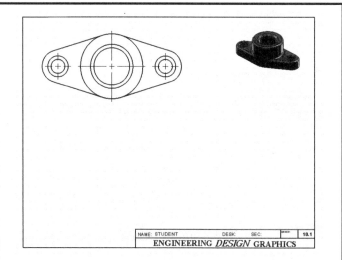

Figure 10-27. Projecting Top View of the Pipe Joint onto a Drawing Sheet.

Now pull down **Window** and select **Tile Vertically**. You now can see the two files: the PIPE JOINT part file and the newly created blank Drawing Sheet file. Now go to **Insert – Drawing View – Model** and select the **Top View**. If SOLIDWORKS does not automatically project a top orthographic layout using the third angle projection you set, as displayed in **Figure 10-27** select the top view in the feature manager tree. *Maximize* the drawing sheet Notice that the view is placed properly.

Pull down the **Insert** menu, select **Drawing View**, and then pick **Section**. Do not forget to change the **Cutting Plane** line thickness to **0.35mm**. A menu will appear with the Horizontal Section line showing as the default. Move your cursor toward the center of the Top View. The cutting Plane Line will snap to the center. The Section Menu will ask you to Select OK. When you do, drag your cursor downward to the place where you want to position your Section view. Notice the vertical alignment between the two views is maintained as you move the section view. Also, a cutting plane line labeled **A-A** appears where you placed the Section Line.

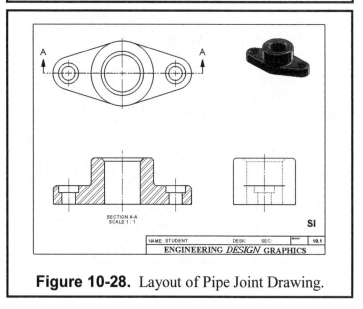

Figure 10-28. Layout of Pipe Joint Drawing.

Next, select **Insert – Drawing View – Projected View** and insert the right side view to the right of the section view. After inserting centerlines, center marks and the small pictorial of the model, your drawing should look like the image in **Figure 10-28**.

SETTING THE DIMENSION VARIABLES

You are now ready to start dimensioning the **PIPE JOINT** model. But first you should set some general dimensioning variables. Pull-down **Tools**, and select **Options**. Then pick the **Document Properties** tab and select the **Dimensions** menu. This opens the menu, as shown earlier in **Figure 10-5**. Click **off** the "Add parentheses by default" variable so that the dimensions do not have parentheses around them. Next, select the **Annotations Font** tab on the **Document Properties** menu. Click **Dimension** and the "Choose Font" menu appears. Next click the **Font** button on the same menu. The "Choose Font" pop-up menu appears. Here you can set the following values:

Font = **Arial**
Font Style = **Regular**
Height = **12 Points or 3mm**.

While in this menu make the
following changes.
Bent Leaders – change to **3 mm**.
Arrows – change values to **1mm, 3mm and 6mm** respectively.
Then click **OK** to close the "Dimension" menu.

DIMENSIONING THE DRAWING

You will start dimensioning the top view. Click on the **Top** view and a blue border appears around it. Now click the **Dimension** tool icon. Pick the *diameter* of the large hole on the right side and drag the **φ18** value outside the object and position it as shown in **Figure 10-29**. Next, pick the rounded end and drag the **R14** *radius* over as shown in **Figure 10-29**. When you place these two dimensions a text box appears to the left of the screen labeled **DIMENSION TEXT**. You can add the text of **2 X** for the **R14** and **2 X** plus the information necessary for a **Counterbore Note**, **Figure 10-29**.

Next, dimension the **φ32** *diameter* of the through hole on top of the boss. *Recall:* If the dimension arrowhead styles are not the way you want them, you can always right mouse click (**RMB**) over the dimension value and select **Properties**. Then use the "Dimension Properties" menu to set your arrowhead style (see earlier **Figure 10-17**), or activate the dimension in the edit mode so there are cyan dots on the arrows. Click on these cyan dots and the arrows will move to the other option. Finally, dimension the **48 mm** distance from the center of the boss to the center of a bolt hole and then one **96mm** dimension between the centers of the bolt holes.

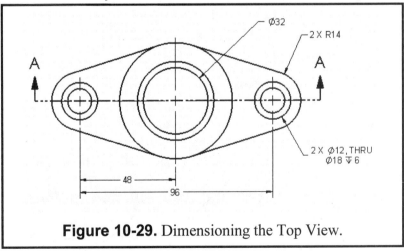

Figure 10-29. Dimensioning the Top View.

Remember that clicking on the 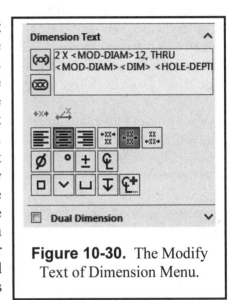 icon will automatically place your dimensions for you. This finishes the dimensioning of the Top View as shown in **Figure 10-29**.

Next, you will dimension the front view. Click on the **Front** view and a *blue* border appears around it. Now click the **Dimension** tool icon. The diameter of the large boss appears in the top view, but dimensioning standards require that you dimension it in the cylindrical view. So pick the length of the boss (that represents its diameter) in the front view and drag the **56** mm value to an appropriate position. This is a diameter value, but the φ symbol does not automatically appear as it did earlier for the top view diameters. You can add this symbol. While you are dimensioning there is a Dimension Window open in the Feature manager Tree. Toward the bottom of this Menu you will see a box labeled **Dimension Text**. Place your cursor just in front of the <dim>, **Figure 10-30** then go down and select the Diameter symbol. You will see a preview of this on your drawing where you placed the original dimension.

Figure 10-30. The Modify Text of Dimension Menu.

Finally, apply the two height **Dimensions** on the right side of the front view. These two dimensions may require some skill. First select the *bottom right corner* of the base length. A dimension may quickly pop up with just this one selection, but disregard it. Next pick the *top surface* of the base plate, and the correct **16** mm height dimension should now appear. Drag it over to an appropriate place on the drawing. Return and pick the *top right corner* of the boss length, and this time select the *bottom surface* of the base plate. The correct **40** mm height dimension should now appear. Drag it over to an appropriate place on the drawing. The front view dimensioning is now complete as shown in **Figure 10-27**.

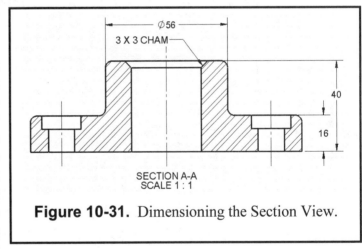

Figure 10-31. Dimensioning the Section View.

You now need to add some centerlines in both the top and front views. Use **Insert, Annotations, Centerline** and click on the box around the front view. The centerlines will be added automatically. You can use **Insert, Annotations** and **Center Mark** to get the centerlines in the top view. Select the two arcs on the ends and the large circle in the center.

Now add a shaded pictorial image to the drawing. Right click the mouse (**RMB**) on the drawing sheet and the pop-up menu (**Figure 10-12**) appears. Select **Drawing View**, then **Model**. On the "Model View" menu (**Figure 10-13**) select **Trimetric** orientation, **Shaded** style, and

Scale of **1:2**. Move the cursor to the upper right corner and insert the pictorial. Right click on the isometric viewport and select tangent edge and turn on tangent lines.

Next use **Insert, Annotations, Note** for adding a **PIPE JOINT** label and **SCALE 1:1**. Use **Arial** style **22** point and **12** point respectively. Add the **SI** (metric) symbol in the bottom right corner. Use **Arial** style and **28** point for this symbol. Also add a note for fillets and rounds: **ALL FILLETS & ROUNDS: R3** using **12** point **Arial**. Add your name and class data to this Title Block Drawing Sheet using **Arial** style and **12** point.

Note: You can move the dimensioned views over a little to the left if you need more room on the right side of the drawing sheet for all these annotations. Just pick the border of the front view (it turns cyan) and **Move** it. The top view will automatically follow since it is aligned with it. Your drawing should now be complete, as suggested in **Figure 10-28**. Now **Save** your drawing and name it **PIPE JOINT.slddrw**. Then **Print** a hard copy to submit to your instructor before you leave the computer lab.

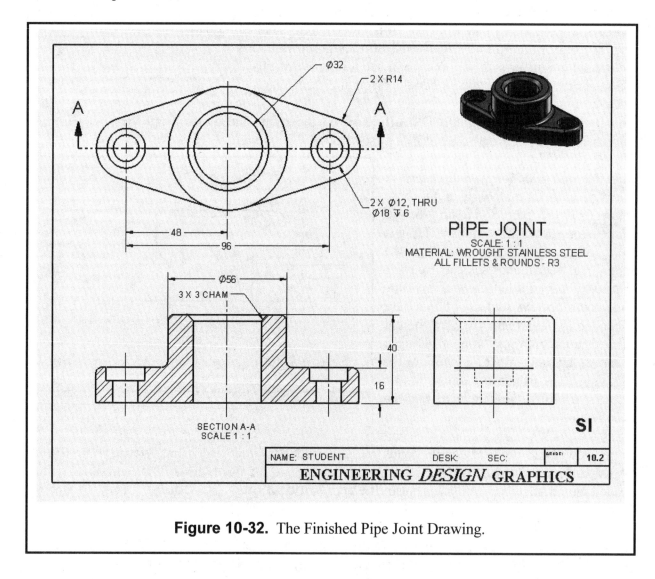

Figure 10-32. The Finished Pipe Joint Drawing.

Exercise 10.3: PEDESTAL BASE Drawing

Quite often in mechanical design, features such as holes are repeated in a regular circular pattern. Examples include flanged pipe joints, inspection cover plates, portal bezels, motor end housings, and many others. Circular and rectangular patterns have been introduced in earlier exercises; here, we create a part with multiple bolt circle patterns and discuss dimensioning practices for the part in a multi-view drawing. You will create a Pedestal Base typical of a stage in a precision manufacturing tool.

CONSTRUCTING THE MODEL

Open **ANSI-METRIC.prtdot** file in SOLIDWORKS. Immediately do a **Save As - PEDESTAL BASE.sldprt**.

Next, select the **Front Plane** in the **Feature Manager** and choose a **Normal** view. Enter the **Sketch** mode and create the sketch for the **Base Revolve**, as shown in **Figure 10-33**. Control the sketch with dimensions as shown, and insert a vertical centerline through the origin. When the profile is complete, select all line segments and the centerline and go to **Insert/Base/Revolve**. Then accept the default condition in the dialog box, which is to create a **360-degree** revolve. Select √ to complete this step. The resulting part will look like **Figure 10-34** when viewed in **Trimetric** mode.

Next, create the circular hole patterns in the pedestal base. There are two primary methods for doing this: you could use the **Insert/Feature/Hole Wizard** to both specify and place the base hole in each pattern, or you can use a sketch in the appropriate plane as the basis for an **Insert/Cut Extrude** to create the holes. The **Hole Wizard** is well suited to more complex features such as counterbored holes (see the next Exercise 10.4), so you will use an **Extruded Cut** in this exercise. You should feel comfortable using either method once Lab 10 is completed.

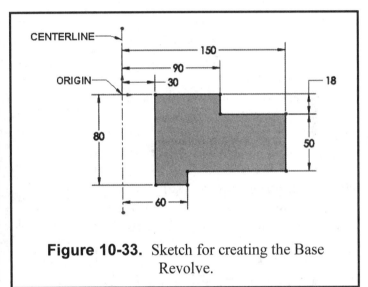

Figure 10-33. Sketch for creating the Base Revolve.

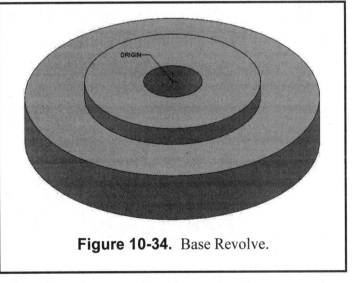

Figure 10-34. Base Revolve.

Select the outer top surface of the **Part**, choose a **Normal** view, and enter the **Sketch** mode. **Draw two Centerlines and a Construction Circle** as elements, placed and dimensioned as shown in **Figure 10-35**. When drawing the 240 mm diameter circle, change the circle to "For Construction" in the Feature Manager Tree. Use these features and the dimension tool to **Draw two Circles** at the intersections, sized at φ**18** mm and φ**8** mm, as shown. Next, use the **Circular Sketch Pattern** tool to create **8** total instances of each size hole, by designating the **Radius** as **120 mm**, the **X** and **Y** coordinates of the center to be **0, 0**, and the **Total Angle** as **360 Degrees**. Accept these values and the repeated holes in each circular pattern will appear as in **Figure 10-36** (they will also be previewed before you close the dialog box). While still in the **Sketch** mode, go to **Insert/ Cut/Extrude** and choose the **Through All** method.

With a **5mm** radius, **Fillet** all outer edges and intersections. Your completed model should look like **Figure 10-37** in the **Trimetric** mode. **Right Click** on "**Material, not specified>**" and **Edit Material**. Expand the Copper Alloys directory and **Select Copper**. If you do not want your Pedestal to take on the color of the material, pick the Appearance Tab and uncheck "Apply appearance of: Copper."

Save your part **File** as **PEDESTAL BASE.sldprt**.

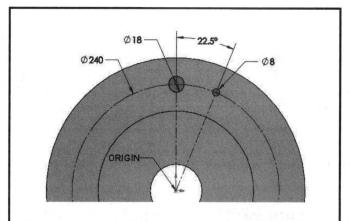

Figure 10-35. Dimensioned Construction Lines.

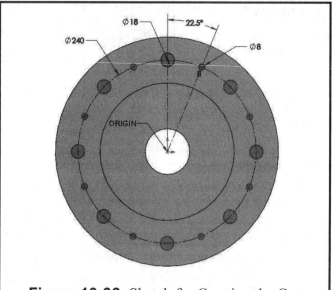

Figure 10-36. Sketch for Creating the Cut Extrude.

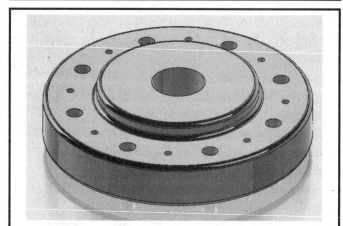

Figure 10-37. Completed Pedestal Base Model.

CREATING THE DRAWING

Now **Open** your **TITLEBLOCK-METRIC.drwdot** drawing sheet and **Save** the new file to the same directory as the **Part** file, with the same name but a different extension - **Pedestal Base.slddrw** (they are of different file types so they can be named the same). There are several formatting options to make. Go to **Options** and set the parameters for **Units**, **Decimal** places, and **Grid** as you did earlier for the **Part** file. In addition, go to **Options/Dimensions** and *deselect* **Add Parentheses by Default**. Also, under **Arrows** select a closed and filled arrow style with the following sizes: **H = 1 mm**, **W = 3 mm**, and **L = 6 mm**. Finally, under **Font** choose **Arial** type with size equal to **12** pt. or **3mm**. Accept these changes. There may be more tweaking required later depending on existing default parameters.

View both files in **Tile Vertical** mode. Activate the Title Block and go to the **Insert** pull down menu. **Select Drawing View – Model**. The name "PEDESTAL BASE" will appear in the **Open Documents** box. Click on the **White Circle with the Blue arrow** in it. In the orientation Box **Click** the box next to **Create Multiple Views**. The menu, where the Feature Manager Tree normally is, will prompt you to identify the views you wish to project. Select the **Top view** and the **Trimetric View**. Select the **Blue Dashed box** around the **Trimetric View**. Make sure that the **Scale** is set to **1:4**. Move the view into the upper right-hand corner of the title block. Now insert a front view as a Full Section by **Inserting – Drawing View – Section View** and also **Insert Drawing view – Projected view** to produce a Right Side View; see **Figure 10-38**.

To place the centerlines in the front view, use **Insert, Annotations** and **Centerline**. Next, activate the **Front View** and the centerlines will be added. Use **Insert, Annotations** and **Center Mark** to get the centerline in the top view as shown in **Figure 10-38**. Also place an *angled* **Centerline** as shown with respect to the vertical centerline in the top view—this dimensioned centerline establishes the "clocking" relationship between the two circular bolt hole patterns. Also in the top view, **Draw** a **Circle** from the center of the object through the center of one of the bolt holes, then **Check "For Construction"** indicating the location of the two circular arrays (bolt circles).

Now, use the **Dimension** tool in the drawing file to completely dimension the drawing. First, use appropriate datums and baseline dimensions to size the front view profile. Also in this view add the outer diameter dimensions for the round features—use the **Modify** box in the dimension dialog to add the appropriate *diameter symbol* ϕ. In the top view, insert diameter dimensions for each of the two sized holes in the respective circular patterns, and each time add the additional information in the **DIMENSION TEXT** box to include the specification of **EIGHT HOLES, EQUALLY SPACED, ON A 240 MM DIAMETER BOLT CIRCLE** (see the notation used in **Figure 10-38**). Add a diameter dimension in the top view for the inner through hole as well.

Dimensioning solutions and styles vary with company practices and drawing standards applied, but your completed drawing should resemble **Figure 10-38**. Since it is a metric drawing, you can add the **SI** symbol. Also, insert a **Trimetric** pictorial in the upper right corner of the drawing. When you have done any necessary modifications to your dimension properties, **Save** all work and **Print** a hard copy for your instructor.

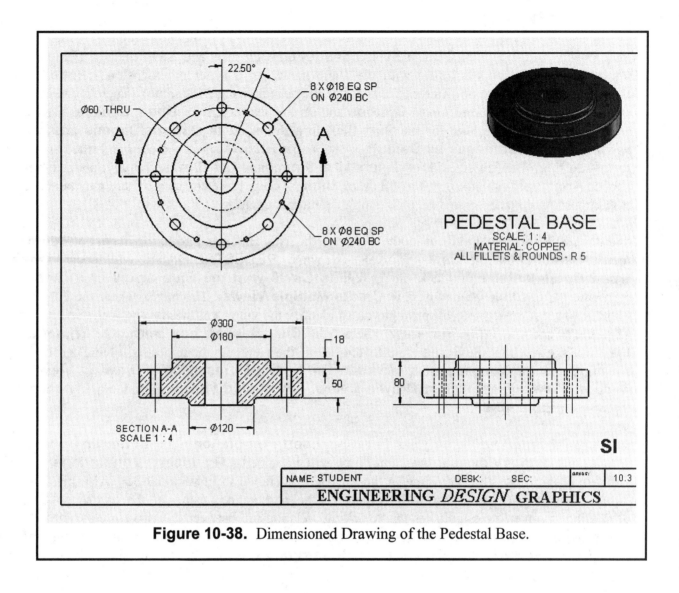

Figure 10-38. Dimensioned Drawing of the Pedestal Base.

Exercise 10.4: TOOLING PAD Drawing

In Exercise 10.4, you will build a solid model of the Tooling Pad using SOLIDWORKS commands that you have learned in previous labs. Next, you will insert the model onto a drawing sheet and then create a three-view orthographic drawing of the part. You will add dimensions and complete the drawing with annotations on a Title block.

BUILDING THE TOOLING PAD MODEL

Open **ANSI-INCHES.prtdot** and Save As – **TOOLING PAD.sldprt**.

Select the **Top** plane and use a **Top** view orientation. In the **Sketch** mode, use the **Rectangle** tool to draw the outline as shown in **Figure 10-35**. Use the **Dimension** tool to apply the given dimensions. Then **Extrude** the profile *upward* from the plane using a **Blind** end condition of **0.725** inches.

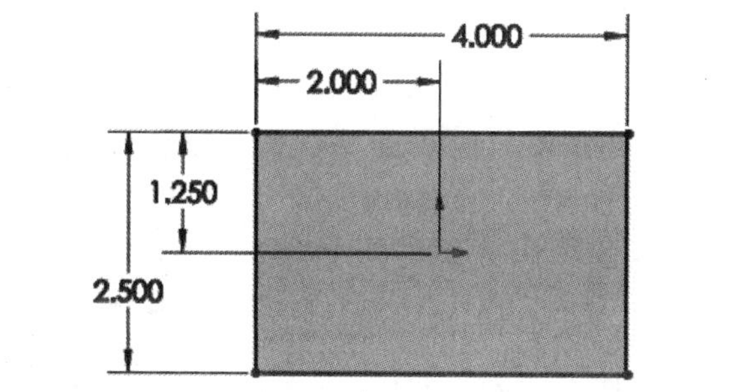

Figure 10-39. Initial Sketch for the Tooling Pad.

Adding a Hole

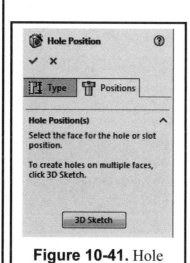

Now you will use the Hole Wizard to place some holes on the Tooling Pad. Click on the top surface of the base part (it turns *blue*) and then pull down **Insert, Features, Hole Wizard**. The "Hole Definition" menu appears, as indicated in **Figure 10-40**. Click on the "Hole" tab and make the following settings:

Standard: **Ansi Inch**
Type: **Fractional Drill Sizes**
End Condition: **Through All**
Hole Diameter: **3/8 in**

Then click the **POSITIONS** tab. This brings you to the second stage of the Wizard as shown in **Figure 10-41**. Select two more points on the surface for the other two holes. It will display faint images of the holes' positions and ask you to use dimensions to position the centers of the holes. Apply the locating dimensions as shown in **Figure 10-42** and okay it.

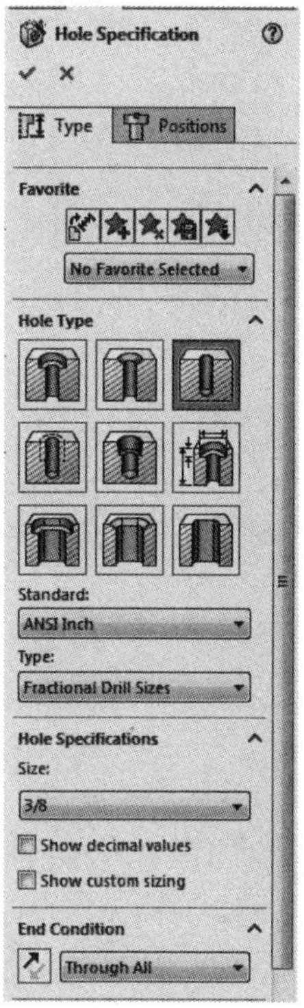

Figure 10-40. The Hole Wizard Menu.

Figure 10-41. Hole Position.

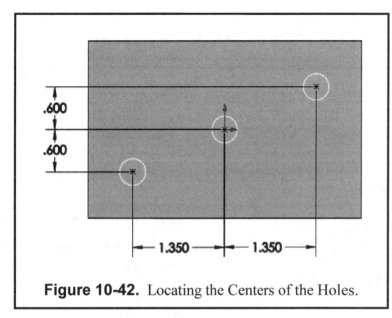

Figure 10-42. Locating the Centers of the Holes.

Adding a Counterbore

Now you will add a counterbore hole using the same "Hole Wizard." Click on the top surface of the base part (it turns *blue*). Pull down **Insert, Features, Hole**, and then **Wizard**. The "Hole Definition" menu appears, as indicated in **Figure 10-43**. Click on the "Counterbore" tab and make the following settings:

> Standard = **Ansi Inch**
> Type = **Hex Bolt**
> Size = **3/8 in**
> End Condition = **Through All**

Select **Show Custom Sizing** and you will see the defaults for this size bolt. Make sure that all the boxes under **Options** are unselected.

Now select the **Positions** tab. This brings you to the second stage of the Wizard as shown in **Figure 10-44**. It will display faint images of the counterbores and ask you to use the **Dimension** tool to locate the center of the faint counterbore holes that appear. Apply the locating

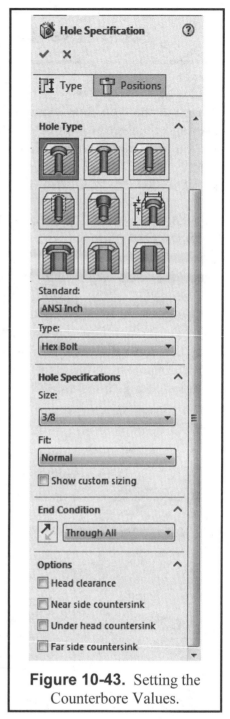

Figure 10-43. Setting the Counterbore Values.

dimensions as shown in **Figure 10-44** and then click the okay button on the pop-up menu. Basically, the counterbore is **.750** inches vertically and horizontally from the *upper left corner* of the base plate. Add a small chamfer on the upper edges of the Counterbore Hole. **Select** the **Feature – Chamfer** and enter the Distance of **.02**.

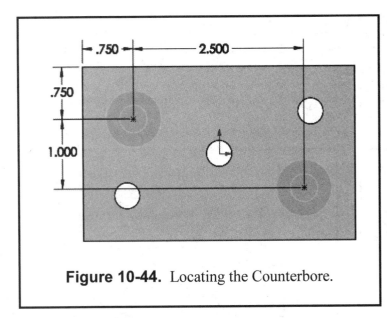

Figure 10-44. Locating the Counterbore.

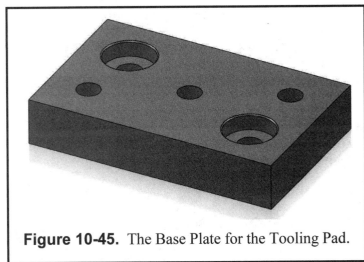

Figure 10-45. The Base Plate for the Tooling Pad.

Adding a Countersink

Now you will add a countersink hole using the same "Hole Wizard." Click on the front surface of the base part (it turns *cyan*). Pull down **Insert, Features, Hole**, and then **Wizard**. The "Hole Definition" menu appears, as indicated in **Figure 10-46**. Click on the "Countersink" tab and make the following settings:

Standard = **Ansi Inch**
Type = **Flat Head Screw**
Size = **.25 in**
End Condition = **Up to Surface**

*For the "End Condition" setting, click the *inner surface* of the small middle through hole.

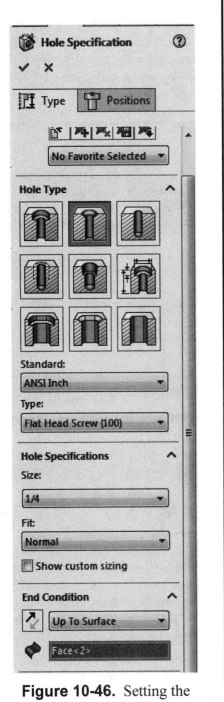

Figure 10-46. Setting the Countersink Values.

Click the **Position Tab**. This brings you to the second stage of the Wizard. It will display a faint image of a countersink and ask you to use the **Dimension** tool to locate the center of the faint countersunk hole that appears. Apply two dimensions to locate it on the front surface: *horizontal* from the left edge = **2.00** inches and *vertical* from the top edge = **.325** inches. Then okay the menu settings to apply the countersink. An **Isometric** view of the model at this stage is shown in **Figure 10-47**.

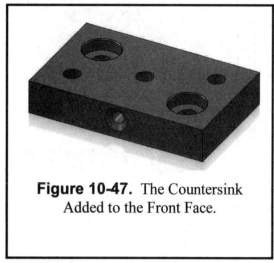

Figure 10-47. The Countersink Added to the Front Face.

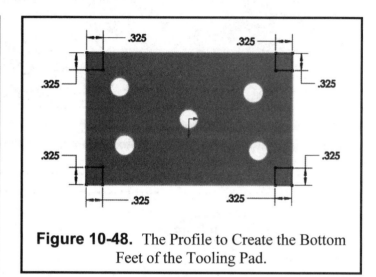

Figure 10-48. The Profile to Create the Bottom Feet of the Tooling Pad.

Adding the Bottom Base Feet

Switch to a **Bottom** view to produce the four corner feet to the Tooling Pad. Use the **Sketch** tools to draw the profile as shown in **Figure 10-48**. Draw four small **Rectangles** in the four corners, and then use add the **.325 Dimensions** on the feet. Next, **Extrude** the sketch **.125** inches **Blind** away from the model. Add **.0625 Fillets** to the outer top edges and the outer vertical edges of the tooling pad. Right mouse click on "**Material <not specified>**," **Edit Material**, then **Expand** the **Titanium Alloys** to **Select** "**Titanium Ti-8Mn, Annealed**." Next, choose **Apply** and **Close** to complete the material application.

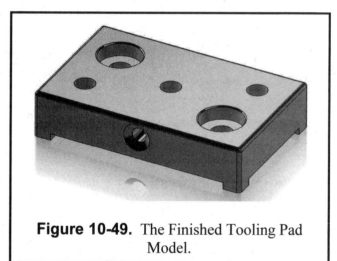

Figure 10-49. The Finished Tooling Pad Model.

The finished model is now shown in **Figure 10-49** in a **Trimetric** view. Use the **Rotate** button to view your various features of the finished model, if you wish. Then **Save** it as **TOOLING PAD.sldprt** in your designated folder, but *do not Close* your file. You can now make a three-view projection.

PROJECTING THREE VIEWS

Pull down **File** and select the **TITLEBLOCK-INCHES** Drawing Sheet. Immediately **SAVE AS TOOLING PAD.slddrw**.

Before you start the projection, there are some other general drawing sheet settings to make (refer back to **Figure 10-8**). Pull down the **Tools** menu and select **Options**. Click the **System Options – Drawings - Display Type** tab. Check the dot (•) on the "Hidden lines visible" option and check the dot (•) on the tangent edges "Removed" option. Also, on the **Document Properties**, **Detailing** tab, set "Dimensioning Standard" to **ANSI**. Click **OK** to close the menu.

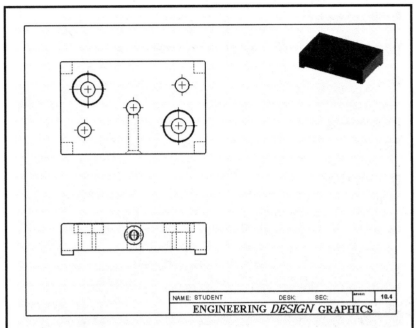

View both files in **Tile Vertical** mode. Activate the Title Block and go to the **Insert** pull down menu. **Select Drawing View – Model**. The name "TOOLING PAD" will appear in the **Open Documents** box. Click on the **White Circle with the Blue arrow** in it. In the orientation box **Click** the box next to "**Create Multiple Views**." The menu, where the Feature Manager Tree normally is, will prompt you to identify the views you wish to project. Select the **Top view**, **Front View**, and the **Trimetric View**. Select the **Blue Dashed box** around the

Figure 10-50. Projecting Two Views of the Tooling Pad onto a Drawing Sheet.

Trimetric View. Make sure that the **Scale** is set to **1:2**. Move the view into the upper right-hand corner of the title block. Select the **Blue Dashed box** around the **Front View**, and set the "Scale" to **1:1**, and then close the menu. See **Figure 10-50.**

CREATING A SECTION VIEW FOR THE RIGHT SIDE

You will create a section view for the right side view to better show the countersink hole.

Pull down the **Insert** menu, select **Drawing View**, and then pick **Section**. A menu will appear; select the Vertical Section line then move your cursor toward the countersink hole in the center of the Front View. The cutting Plane Line will snap to the center. The Section Menu will ask you to Select OK. Now drag your cursor toward the right of the front view and place it where you want to position your Section view. Turn off the hidden lines if they are showing in the section view. See **Figure 10-51.**

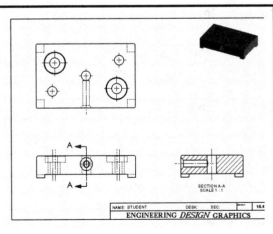

Figure 10-51. Creating a Section View of the Tooling Pad.

ADDING CENTERLINES

The drawing may not be perfect. For example, the centerlines in the views may be missing or incomplete. So **INSERT, ANNOTATIONS, Centerline** and activate each of the views to automatically add the centerlines in the Top, Front, and Section views. For any Center marks missing, **Insert**, **Annotations**, and select **Center Mark**. Then click on each perimeter of the circular features to add the center marks (i.e., centerlines).

SETTING THE DIMENSION VARIABLES

You are now ready to start dimensioning the Tooling Pad drawing. But first set some general dimensioning variables. Pull-down **Tools**, and select **Options**. Then pick the **Document Properties** tab and select the **Dimensions** menu. This opens the menu, as shown earlier in **Figure 10-8, page 10-4**. Click **off** the "Add parentheses by default" variable so that the dimensions do not have parentheses around them. Next, click the **Annotations Font, Dimension** option on the same menu. The "Choose Font" pop-up menu appears, as shown earlier in **Figure 10-9**. Here you can set the following values:

Font = **Arial**
Font Style = **Regular**
Height = **12 Points**.

While in this menu make the following changes.
Bent Leaders – change to **.125** inches.
Arrows – change values to **.042, .125 and .25** respectively.
Then click **OK** to close the "Choose Font" menu. Next, click on the **Units** tab and set the units to **3** decimal place **Inches**, then click **OK** to close the menu.

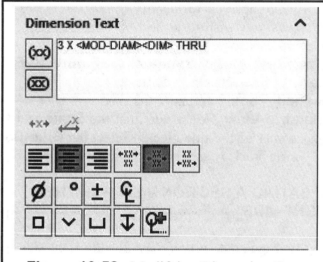

Figure 10-52. Modifying Dimension Text.

DIMENSIONING THE DRAWING

You will start dimensioning the top view. Click on the **Top** view and a cyan border appears around it. Now click the **Dimension** tool icon. Pick the *diameter* of the small hole in the upper right corner and drag the φ.375 value outside the object and position it as shown later in **Figure 10-53**. Go to the **Dimension Text** box in the **Feature Manager** tree and add **3 X** in front of the default dimension and the word **THRU** after the dimension to cover the other two small holes (See **Figure 10-52**).

Next, **Dimension** the position of the hole relative to the upper right corner (**.650** in both directions). Add the note **TYP** after each of the dimensions. Go to the **Dimension Text** box as shown in **Figure 10-53** and type in **TYP** after the <DIM> to add this note to the dimension value. Now dimension the counterbore. Click the **Dimension** tool icon and pick the *outer diameter* of the large counterbore **(Not the Chamfer Circle)** in the upper left corner. Drag the

φ.875 value outside the object and position it as shown in **Figure 10-53**. Notice that the method of dimensioning a counterbore is to call out the small thru hole first. So repeat the previous "Modify Text of Dimension" process (see **Figure 10-49**) to change the dimension value to **φ.375, THRU**. Now below it, add the rest of the call out using an **Annotations Note**:

2 X ⌀.375 THRU
⌴ **⌀.875 ▽.25**

The symbol <MOD-DIAM> just adds the φ symbol in front of the **0.875** diameter for the counterbore. Now complete the linear dimensions of the top view as suggested by **Figure 10-53**. You can **Dimension** the **2.500** depth and the **4.000** width. You can position the counterbore as **0.750 TYP** from the bottom right corner in both directions. You can position the small center hole several ways, one of which is depicted below.

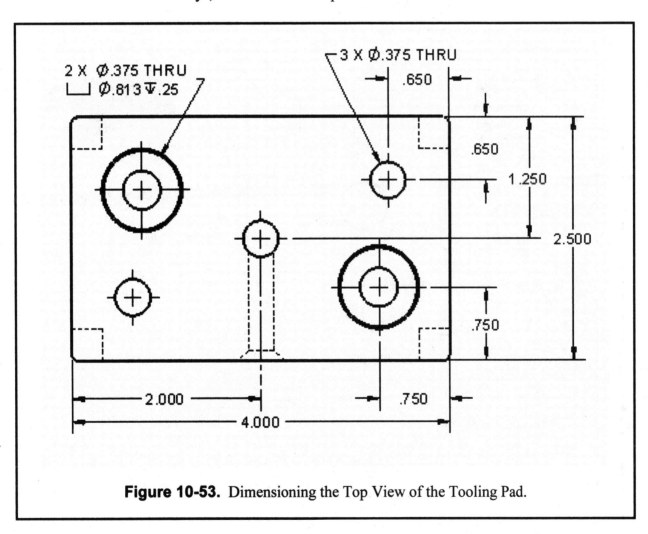

Figure 10-53. Dimensioning the Top View of the Tooling Pad.

Now start your dimensioning on the front view. Make a call out for the countersink using a similar technique as the counterbore. Make the note as follows:

$$\varnothing.250 \,\overline{\underline{\vee}}\, 1.25$$
$$\vee \,\varnothing.507 \times 82°$$

Complete the exercise by adding the remaining linear dimensions to the front and right side section view, as suggested by the finished drawing in **Figure 10-54**. Use your choice of style.

Next use **Insert, Annotations, Note** for adding a **TOOLING PAD** label and **SCALE 1=1**. Use **Arial** style **22** point and **12** point respectively. Add your name and class data inside the title block using **Arial** style and **12** point. Your drawing should now be complete, as shown in **Figure 10-54**. Now **Save** your drawing and name it **Tooling Pad**. Then **Print** a hard copy to submit to your instructor before you leave the computer lab.

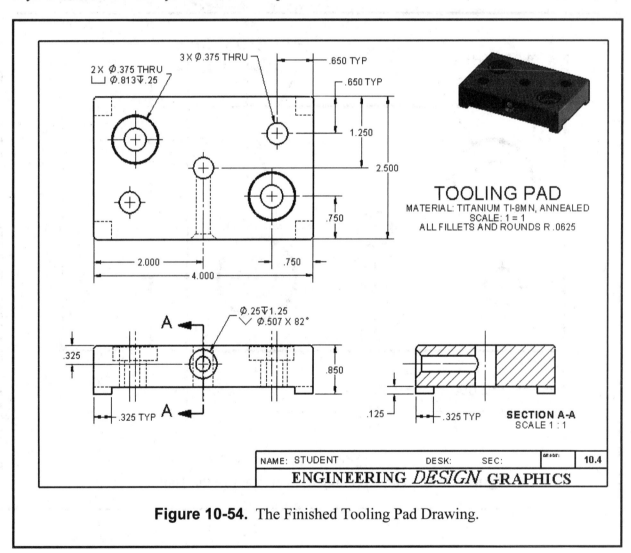

Figure 10-54. The Finished Tooling Pad Drawing.

SUPPLEMENTARY EXERCISE 10-5: PILLOW BLOCK

Build a full sized solid model of the figure below. Insert it on a Title Block and Dimension it. Insert a small pictorial of the object in the upper right hand corner of the sheet. Provide the proper Titles, Scales and other pertinent notes.

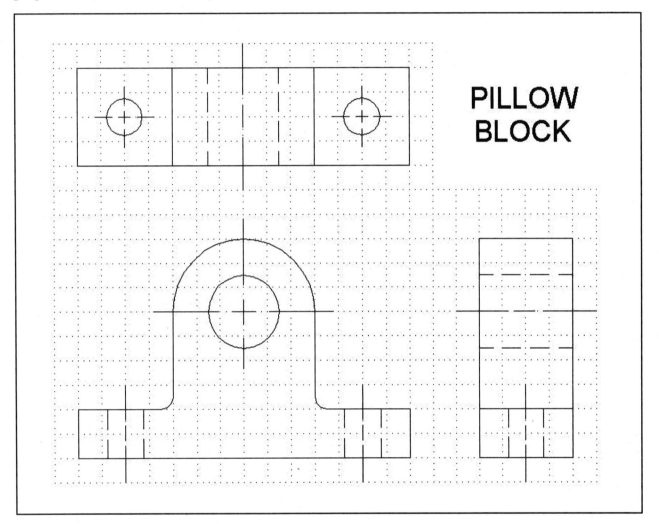

PILLOW BLOCK

ASSUME THE GRID DIVISIONS TO BE 5 mm.

SUPPLEMENTARY EXERCISE 10-6: CHASSIS BOX

Build a full sized solid model of the figure below. Insert it on a Title Block. Dimension the drawing according to standards learned doing the exercises in Unit 10. Finally, insert a small Isometric View in the upper right hand corner. Provide the proper Titles, Scales and other pertinent notes.

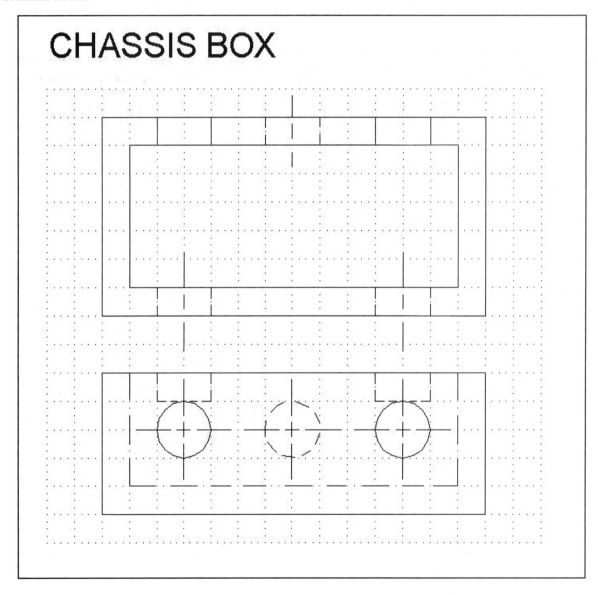

ASSUME THE GRID DIVISIONS TO BE 0.50 INCHES.

SUPPLEMENTARY EXERCISE 10-7: COLUMN BASE

Build a full sized solid model of the **COLUMN BASE** below. Insert it on a Title Block. After deleting the front and the right side view, draw a horizontal line through center of the top view. Next, go to Insert – Drawing View – Section View and place it where the front view used to be. Dimension the drawing according to the standards learned in Unit 10. Finally, insert a small Isometric View in the upper right hand corner. Provide the proper Titles, Scales and other pertinent notes.

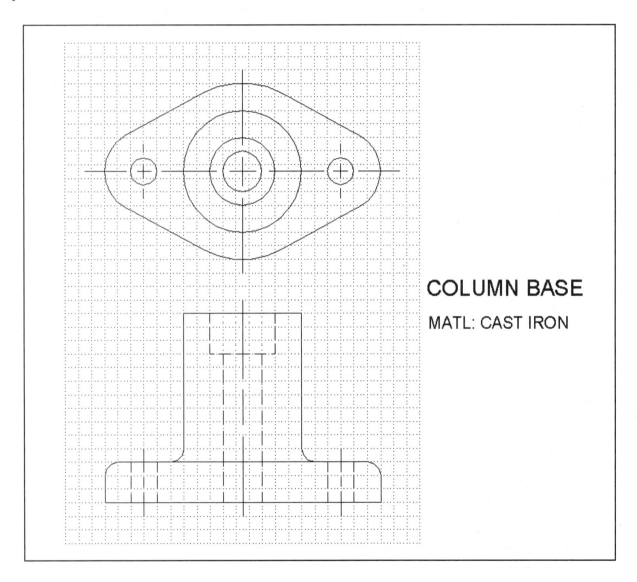

COLUMN BASE

MATL: CAST IRON

ASSUME THE GRID DIVISIONS TO BE 0.25 INCHES.

SUPPLEMENTARY EXERCISE 10-8: VALVE HOUSING

Build full sized solid models of the **VALVE HOUSING** below. After deleting the front and the right side view, draw a horizontal line through center of the top view. Next go to Insert – Drawing View – Section View and place it where the front view used to be. Also Insert – Drawing View – Projected View to replace the Right Side view. Dimension the drawing according to the standards learned in Unit 10. Finally, insert a small Isometric View in the upper right hand corner. Provide the proper Titles, Scales and other pertinent notes.

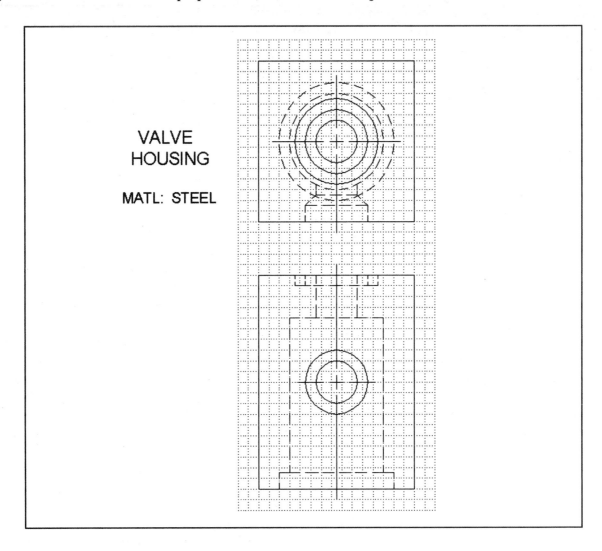

VALVE
HOUSING

MATL: STEEL

ASSUME THE GRID DIVISIONS TO BE 0.25 INCHES.

APPENDIX A
EXAMPLE OF A TILEBLOCK WITH DIMENSIONS

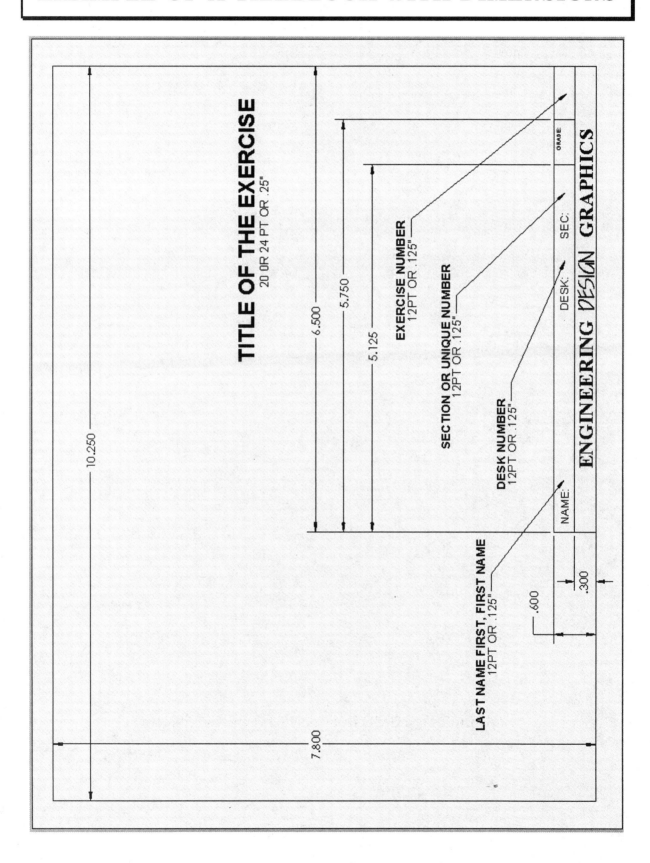

NOTES:

NOTES:

NOTES:

NOTES: